Biologics Development: A Regulatory Overview

Biologics Development: A Regulatory Overview

Edited by
Mark Mathieu

PAREXEL International Corporation
Waltham, MA
Publishers

Biologics Development: A Regulatory Overview

Edited by
Mark Mathieu
PAREXEL International Corporation

Mark D. Harris:	*Design and Production*
Joanna White:	*Design and Production*
James Applebaum:	*Cover Design*
Carolyn Newman:	*Marketing*
Victoria Wong:	*Fulfillment*

Library of Congress Catalog Card Number 92-62597

Acknowledgements

My sincerest thanks to the individuals who took time from their own demanding schedules to aid my efforts: Erez Azaria, Duncan Berkeley, John Caynon, Rick Davis, Cori Doud, Veronica Emery, Adrienne Garland, Robert Goldstein, Alberto Grignolo, Ph.D., Mark Harris, David Kaplan, John Kirchner, Wayne Kubick, Lisa L'Hote, Kaitlin MacDonald, Carolyn Newman, Seth Pauker, Ph.D., Robert Regazzi, Anne Sayigh, Ph.D., Halpreet Sing, Thomas Steen, Cheryl Stefanelli, Marni Sullivan, Josef von Rickenbach, Anne Whitaker, and Victoria Wong.

Special thanks to my contributors, particularly those within the Center for Biologics Evaluation and Research (CBER), who worked diligently on this project through an extremely trying reorganization period. Also, special thanks to the Center for the Study of Drug Development at Tufts University, particularly Drusilla Raiford.

Preface

In the span of several weeks in late 1992, the manner in which biologics and biotechnology products are reviewed and approved by the U.S. Food and Drug Administration (FDA) took a fundamental turn. First, the government approved an FDA reorganization plan that, in essence, "re-engineered" the review and approval process for these products. Just as important, President Bush signed legislation that permits the FDA to impose fees on sponsors of drug and biologics products, fees that the agency promised to use to halve product review times by October 1997.

The target of these changes—the FDA's Center for Biologics Evaluation and Research (CBER)—had quickly taken on increased importance beyond its role in protecting American consumers. As the FDA unit that regulates biologics and most biotechnology products, CBER's ability to handle the influx of research and marketing applications had become a political and economic issue as well as a health-care issue. The federal government had become increasingly concerned about the health of the fledgling U.S. biotechnology industry, which most world governments also recognized as an industry critical to competitiveness in the twenty-first century.

To bring the speedier approvals so essential to the biotechnology industry, the U.S. government took steps to fortify CBER staffing and resources. Under the Prescription Drug User Fee Act of 1992, CBER is to receive resources to fund 300 new product review-related positions over the next several years. In turn, CBER reorganized not only to absorb the new staffing but to streamline the review process as well.

Such rapid and fundamental change brings additional uncertainty to a biological product approval process that was, at the start, not necessarily well understood or documented. There are several real and identifiable reasons for this. First, the products that CBER regulates are growing in complexity, and are far less homogeneous than products regulated by CBER's sister centers within the agency. Just as important, the majority of products now funneling into CBER originate from the biotechnology industry, an industry with which the FDA has only a decade of experience and many of whose companies are just now addressing regulatory issues.

This text represents the first attempt to document the FDA's biological product approval process at such a level of detail. And following a period of such fundamental change, the book will represent, for many, an introduction to the "new" CBER and a new era of biologic regulation. This change, which has made this text such a challenge to develop, is also likely to be what makes the volume particularly valuable.

That stated, it is important to point out that CBER and its new review processes will continue to evolve rather quickly. Wherever possible, the authors have identified areas known to be targeted for change, and have attempted to characterize the nature of the expected change.

It is also worth noting that the considerable diversity of biologics has resulted in equally diverse development and regulatory processes for biological products. The FDA's approval standards and processes as they apply to two biological products may be vastly different. Therefore, it is virtually impossible to provide all-encompassing descriptions of FDA policies, regulatory requirements, and review procedures that are entirely relevant to each new biologic. Inevitably, there are special circumstances that contradict even the most basic outlines of the biologics development process. While this text emphasizes the elements of the development and review process common to all biological products, it also attempts to identify many of the principal differences in the regulatory, testing, and submission requirements for various products.

Finally, although this text features several contributing authors employed by the FDA, it does not necessarily, and does not purport to, represent in all cases the official views of the U.S. FDA or any center within the agency.

Contents

Chapter 1 - Introduction

What is a Biologic?

by Steven R. Scott
Director of Regulatory Affairs
Pharmaceutical Division
Sepracor, Inc.

The United States has not only the world's most productive pharmaceutical industry, but also one of the most sophisticated government systems for ensuring the safety and effectiveness of biomedical products. Today, Americans benefit from living in an era in which scientists are quickly unraveling the mysteries of body function and disease interaction, and in which there is a major focus on commercializing this knowledge to improve the quality of human life.

As a world leader in the biomedical sciences (i.e., molecular biology, genetics, immunology, etc.), the U.S. pharmaceutical industry has sponsored biomedical research and produced a new generation of technologies and products intended to prevent and treat human disease. This research will continue to produce novel biomedical products, including drugs, biologics, and medical devices.

In medicine and science, as in other areas, history tends to repeat itself. During the late 1800s and early 1900s, scientific advances led to a cursory understanding of the immune system and specific disease processes. This knowledge resulted in the development of technologies that produced a new class of pharmaceutical agents—biologics.

These early discoveries, coupled with an explosion in scientific advances over the past 25 years, spawned the new biotechnologies of the 1970s, 1980s, and 1990s, and led to the emergence of the biotechnology industry. Academicians, entrepreneurs, and established pharmaceutical companies have fueled this industry, using these technologies in new therapeutic areas

1

and applications. As a result, the past decade has produced a new generation of increasingly complex products intended for use in humans.

In particular, this marriage of science and business has also spawned an increasingly diverse and complex group of products requiring U.S. Food and Drug Administration (FDA) approval. After all, today's biologics are not the comparatively simple products and technologies commonly found 30 years ago.

The diversity and complexity of biotechnology products have contributed greatly to uncertainty over how today's products—drugs and biologics, in particular—should be classified. Such classification issues are extremely important, particularly since there are fundamental differences in the FDA's regulation of drugs and biologics.

To better understand the nature of biological products, it is worth examining the history of biologics regulation, the rationale for the regulatory approaches of federal authorities, and statutory and regulatory definitions for biologics. Doing so provides valuable insights into why biologics are regulated differently than similar products, and how novel, cutting-edge products are regulated in the United States.

A Brief History of Biologic Regulation in the United States

Biologics and the government's regulation of these products have long and storied histories in the United States. Early in this country's history, epidemics of diphtheria, typhoid, yellow fever, small pox, tuberculosis, and cholera were rampant. As scientists began to understand the causes and symptoms of these and other diseases, medical advances quickly followed. The work of Jenner, Pasteur, Koch, Ehrlich, von Behring, and Roux advanced human understanding of the principles of infectious disease, and laid the groundwork for the birth of commercial biologics development.

In the early 1900s, the newborn concepts of vaccination and the therapeutic use of antitoxin led to the founding of American and European laboratories to produce antisera. Since the technology for producing this class of products was new and the concepts of microbiology and immunology were in their infancy, industry had only a rudimentary understanding of how certain manufacturing-related factors affected the safety and effectiveness of these new biologics. Given the lack of government standards for product safety and effectiveness, and the crude analytical and manufacturing methods used at the time, it is not surprising that subpotent and often hazardous biologics were produced and administered to humans.

A tragedy motivated the federal government to regulate biologics. During a serious diphtheria epidemic in 1901, ten children died after being treated with antitoxin produced by the city of St. Louis. An investigation revealed that the antitoxin had been contaminated with tetanus—specifically, the horse from which the antitoxin was obtained had contracted tetanus. No safety testing had been performed on the product before its administration to children.

At that time, the government and medical community raised further concerns, recognizing that commercial ventures would undoubtedly develop antitoxin, and that anyone with some technical knowledge and a stable of horses could produce it. In response to the 1901 tragedy and subsequent concerns about the manufacture of antitoxin, Congress introduced legislation mandating biologics regulation. Approved and signed into law on July 1, 1902, the Biologics Control Act had the following purposes:

- to authorize the regulation of the sale of viruses, serums, toxins, and analogous products;

- to authorize the promulgation of biologics regulations;

- to require licensing of the manufacturing establishment and manufacturer; and

- to provide inspection authority to the federal government.

This was the first federal legislation to regulate a specific class of drugs in the United States. The law made it illegal for companies to transport or sell biologics that were:

- not manufactured at a licensed facility inspected by the federal government; or

- not labeled with the identification of the manufacturer and an expiration date.

3

The statute established a board within the Treasury Department to promulgate regulations for the licensing of companies involved in the manufacture and sale of biologics in interstate or foreign commerce. The board was also given the authority to establish and to maintain standards for the safety, purity, and potency of all biologics falling under the provisions of the law.

In 1903, federal authorities issued the country's first biologics regulations. Administered by the Public Health Service's (PHS) Hygienic Laboratory, these regulations required the annual renewal of establishment licenses and formalized the concept of unannounced government inspections.

Not until 1906 did Congress pass the Pure Food and Drugs Act, which was enacted so federal authorities could regulate other drugs. The law required that drugs not be mislabeled or adulterated, and that they meet the recognized standards of strength and purity. Interestingly, this law did not mention either biologics or the Biologics Control Act of 1902, an omission that, in effect, represented the first distinction between drugs and biologics.

During 1919, the biologics regulations were amended to require that licensed manufacturers report changes in equipment, personnel, and manufacturing methods, and maintain permanent records of manufacturing and quality control procedures. These regulations also required, for the first time, that manufacturers of some products forward product samples for government inspection and approval before releasing batches or lots from which the samples were derived.

In 1930, the Ransdell Act redesignated the Hygienic Laboratory as the National Institutes of Health (NIH), under which its functions were expanded. During 1937, the NIH was reorganized, and the newly established Division of Biologics Control was assigned the responsibility of regulating biologics.

Responding to the sulfanilamide drug tragedy, which took 107 lives, Congress enacted the Food, Drug and Cosmetic Act of 1938 (FD&C Act). This law provided the government with the power to seize drug products that were adulterated or misbranded. The statute is particularly important in the context of this discussion because it also provides the current regulatory definition for "drugs":

> "articles intended for use in the diagnosis, cure, treatment, or prevention of disease in man..." and "articles other than food that affect the structure or any function of the body of man..."

4

Given their nature, the "viruses, serums, toxins, and analogous products" regulated as biologics were also accommodated by this definition. Without stating this fact explicitly, the definition made biologics subject to the provisions of the FD&C Act as well as existing biologics statutes.

In the mid-1940s, when the biologics laws were restated and codified as the Public Health Services Act (PHS Act), the federal government addressed concerns regarding the duplicative regulation of biologics under the FD&C Act and the PHS Act. Since biologics laws specified only premarketing controls and no postmarketing controls for biologics, some in Congress wanted to explicitly state in the PHS Act that all products subject to the law were also subject to the FD&C Act. In particular, they wanted to ensure that the government's power to seize unsafe drug products under the FD&C Act would apply equally to biologics.

However, industry concern over duplicative regulation of biologics forced Congress to relent in its efforts to incorporate such statements in the PHS Act's codification. Instead, Congress opted for a less specific statement that permitted the application of the seizure powers to biologics, but left the licensing controls placed on new biologics and the premarketing controls placed on new drugs separate and intact.

The PHS Act's codification process produced one other change that brought greater similarities to drug and biologics regulation: Congress took this opportunity to extend the government's regulation of the biological industry beyond simply manufacturing establishments to include the product as well. More specifically, this change required that sponsors of new biologics obtain a biological product license in addition to an establishment license.

In 1948, the Division of Biologics Control was made part of the National Microbiological Institute, which later became the Institute of Allergy and Infectious Disease. In 1955, the authority for biologics regulation was transferred to the newly created Division of Biological Standards. This change was prompted, in part, by the distribution of an improperly inactivated lot of poliomyelitis vaccine that resulted in 202 cases of polio, ten of them fatal.

During 1970, biologics laws were amended to include blood, blood components, and allergenic extracts.

The Division of Biological Standards continued to regulate biologics until 1972, when the Food and Drug Administration (FDA) assumed the responsibility. Due to multiple FDA reorganizations, biologics were regulated, for various periods, by the Bureau of Biologics (BoB), the National Center for Drugs and Biologics (NCDB), and the Office of Biologics Research and Review (OBRR). Currently, the FDA's Center for Biologics Evaluation and Research (CBER) regulates biological products (see Chapter 5).

Government Approach to Biologics Regulation

In attempting to understand the FDA's criteria for classifying substances as biologics, and the agency's approach to regulating these products, an analysis of the early biologics (i.e., their content, characterization, and methods of preparation) provides several valuable insights. This is particularly relevant since current regulatory approaches share much with the federal government's attempts to regulate these first biological products.

From the infancy of biologics development to the not-so-distant past, biologics consisted of crude preparations. In the early 1900s, vaccines and antitoxins (e.g., smallpox vaccine, diphtheria, and other antitoxins) were obtained by immunizing animals with an immunogenic challenge to elicit an immune response. The products consisted of crude mixtures of animal sera, which were administered prophylactically or early in the course of disease progression.

Later, advances in the understanding of the human immune system led to the development of prophylactic bacterial vaccines. These products were crude preparations of inactivated forms of the pathogenic organisms or toxins.

During 1941, new manufacturing methods permitted the development of better viral vaccines. Viruses, such as influenza, mumps, yellow fever, polio, measles, and rubella, could be grown in large quantities using eggs or animal tissue cultures—previously, whole animals were needed for viral vaccine production. Killed viral vaccines and attenuated live viral vaccines were developed using these new manufacturing methods, which required the development of appropriate immunogen and vaccine characterizations, manufacturing controls, and product safety testing

In comparing biologics development to conventional, synthetic drug development, the need for a different regulatory approach becomes evident. In the early years of conventional drug development, end-product testing was emphasized in determining a product's performance and safety. Drug substances were either synthetically produced through chemical reactions or isolated and purified from plants or other sources. Scientists could easily determine the purity of the drug substance.

Conversely, researchers could not isolate or identify the active substances in the earliest biologics. Since many of these biologics worked through immune response mechanisms, extremely small quantities of the "active substance" were sufficient to achieve effectiveness. Generally, these products had a purity of less than one percent, with the remainder of the preparation consisting of unwanted proteins and other impurities.

In addition, existing analytical methodologies and protein chemistry techniques were not sensitive enough to adequately characterize the products with respect to their contents. Thus, authorities recognized that end-product testing, while important, was not sufficient to ensure the safety and potency of biologic products.

The methods employed in product manufacturing also represent a significant difference between early biologic and conventional drug development. As stated previously, biological product manufacture required the use of immunized farm animals, which would provide the sera or active substance. Scientists found that product quality and safety could be affected by the health of these animals, the conditions under which they were kept, and the method through which the biologics were collected. Safety hazards regarding the preparation and composition of the inocula used to produce the viruses and toxins were also issues. The tragedy in 1901 substantiated the concern that inoculum could be contaminated with other organisms, resulting in preparations containing pathogenic infectious agents virulent in humans. These issues and concerns were never relevant in conventional drug development or manufacturing.

How did these factors influence the federal government's initial attempts to regulate biologics? Most importantly, they made clear that end-product testing was not adequate to control and characterize biological products. Simply put, the technological limitations of that time did not permit manufacturers to adequately characterize the components and contaminants present in biological preparations through product testing.

Therefore, Congress had to take a different regulatory approach. Instead of emphasizing the biological product, the federal government focused on controlling the product's manufacturing environment (facility) and method of manufacture, since these could have major effects on the quality and safety of the biologic. Even at that time, scientists knew that minor changes in the source of the immunogen (virus or toxin), production vehicles (animals), and processing equipment directly influenced product safety, effectiveness, and quality. Federal authorities saw these as the best—possibly the only—available parameters on which to base regulatory controls for the quality and safety of biologics.

This rationale forms the basis for the licensing provisions in the PHS Act, and provides an understanding of how the current regulatory environment for biologics began its evolution. The focus on the production facilities, method of manufacture, and end-product testing remains intact today, and provides the basis for current approaches to ensure consistency in the manufacture, purity, potency, and safety of biologics.

Regulatory Definition of a Biologic

As stated above, the Biologics Control Act of 1902 was recodified and consolidated in the PHS Act of 1944. Today, this act provides a description of the products under the law's authority, a description that can be used to define biologics:

"[A biologic is] any virus, therapeutic serum, toxin, antitoxin, vaccine, blood, blood component or derivative, allergenic product, or analogous product, or arsphenamine or its derivatives (or any other trivalent organic arsenic compound), applicable to the prevention, treatment, or cure of diseases, or injuries in man ..."

The PHS Act also authorized federal authorities to issue regulations to ensure the safety, purity, and potency of biological preparations. These regulations are codified in Title 21 of the *U.S. Code of Federal Regulations (CFR)*, Parts 600 through 680. Because some provisions of the FD&C Act also apply to biologics, several regulations issued under that law are also relevant to biological products. The table below lists the regulations applicable to biologics development.

REGULATIONS APPLICABLE TO BIOLOGICS		
Regulation	**Title**	**Content**	**Development Impact**
21 CFR Part 25	Environmental Impact Considerations	Environmental Assessment Regulations	Product Manufacture and Licensure
21 CFR Part 50	Protection of Human Subjects	Patient Informed Consent Requirements	Clinical Testing
21 CFR Part 56	Institutional Review Boards	Ethical Review Requirements	Clinical Testing
21 CFR Part 58	Good Laboratory Practices for Nonclinical Testing	Nonclinical Testing Standards	Safety and Toxicology Testing
21 CFR Part 200	Drugs: General	Mailing Requirements for Important Drug Information. Specific Drug Class Requirements	Dissemination of Drug Information
21 CFR Part 201	Labeling	Drug Labeling Requirements	Drug Labeling
21 CFR Part 202	Prescription Drug Advertising	Product Advertising and Promotion Requirements	Promotional Activities
21 CFR Part 210	Current Good Manufacturing Practice in Manufacturing, Processing, Packaging, or Holding of Drugs: General	General Requirements for the Manufacture of Drugs	Product Manufacture

Biologics Development: A Regulatory Overview

REGULATIONS APPLICABLE TO BIOLOGICS (cont'd)

Regulation	Title	Content	Development Impact
21 CFR Part 211	Current Good Manufacturing Practice Requirements for Finished Pharmaceuticals	Requirements for the Manufacture of Finished Pharmaceuticals	Product Manufacture
21 CFR Part 312	Investigational New Drug Application	IND Requirements	Clinical Testing
21 CFR Part 600	Biological Products: General	Administrative and Manufacturing Requirements	Product Manufacture
21 CFR Part 601	Licensing	Product and Facility Licensing Requirements	Product Licensure
21 CFR Part 606	Current Good Manufacturing Practice for Blood and Blood Components	Blood and Blood Component Manufacturing and Processing Requirements	Blood and Blood Component Manufacture
21 CFR Part 610	General Biological Product Standards	Product Testing, Release, and Labeling Requirements	End Product Testing and Labeling
21 CFR Part 620	Additional Standards for Bacterial Products	Standards for Pertussis, Typhoid, Anthrax, Cholera, and BCG Vaccines	Bacterial Product Manufacture, Testing, Labeling and Release
21 CFR Part 630	Additional Standards for Viral Vaccines	Standards for Poliovirus, Measles, Mumps, Rubella, and Smallpox Vaccines	Viral Vaccine Product Manufacture, Testing, Labeling, and Release

REGULATIONS APPLICABLE TO BIOLOGICS (cont'd)

Regulation	Title	Content	Development Impact
21 CFR Part 640	Additional Standards for Human Blood and Blood Products	Standards for Whole Blood, Red Blood Cells, Platelets, Plasma, Cryoprecipitate, Source Plasma, Albumin, Plasma Protein Fraction, Immune Globulin, and Measles Immune Globulin	Blood and Blood Product Manufacturing, Testing, Labeling, and Release
21 CFR Part 650	Additional Standards for Diagnostic Substances for Dermal Tests	Standards for Diphtherial Toxin for Schick Test, Tuberculin Test	Dermal Test Product Manufacturing, Testing, Labeling, and Release
21 CFR Part 660	Additional Standards for Diagnostic Substances for Laboratory Tests	Standards for Antibody to Hepatitis B Surface Antigen, Blood Grouping Reagents, Reagent Red Blood Cells, Hepatitis B Surface Antigen, Anti-Human Globulin and Limulus Amebocyte Lysate	Diagnostic Test Product Manufacturing, Testing, Labeling, and Release
21 CFR Part 680	Additional Standards for Miscellaneous Products	Standards for Allergenic Products, Trivalent Organic Arsenicals, Blood Group Substances	Requirements for Miscellaneous Product Manufacturing, Testing, Labeling, and Release

The biologics regulations are particularly important in the context of this discussion because they provide the most detailed definition of a biologic. Federal regulations, specifically 21 CFR 600.3, state that:

"(h) Biological product means any virus, therapeutic serum, toxin, antitoxin, or analogous product applicable to the prevention, treatment or cure of diseases or injuries in man."

The regulations continue on to define each of the key terms in the definition itself:

"(1) A virus is interpreted to be a product containing the minute living cause of an infectious disease and includes but is not limited to filterable viruses, bacteria, rickettsia, fungi, and protozoa.

(2) A therapeutic serum is a product obtained from blood by removing the clot or clot components and the blood cells.

(3) A toxin is a product containing a soluble substance poisonous to laboratory animals or to man in doses of 1 milliliter or less (or equivalent in weight) of the product, having the property, following the injection of non-fatal doses into an animal, of causing to be produced therein another soluble substance which specifically neutralizes the poisonous substances and which is demonstrable in the serum of the animal thus immunized.

(4) An antitoxin is a product containing the soluble substance in serum or other body fluid of an immunized animal which specifically neutralizes the toxin against which the animal is immune.

(5) A product is analogous:

(i) To a virus if prepared from or with a virus or agent actually or potentially infectious, without regard to the degree of virulence or toxicogenicity of the specific strain used.

(ii) To a therapeutic serum, if composed of whole blood or plasma or containing some organic constituent or product other than a hormone or an amino acid, derived from whole blood, plasma, or a serum.

(iii) To a toxin or antitoxin, if intended, irrespective of its source of origin, to be applicable to the prevention, treatment, or cure of disease or injuries of man through a specific immune process.

(i) Trivalent organic arsenicals means arsphenamine and its derivatives (or any other trivalent organic arsenic compound) applicable to the prevention, treatment, or cure of diseases of man.

(j) A product deemed applicable to the prevention, treatment, or cure of diseases or injuries of man irrespective of the mode of administration or application recommended, including use when intended through administration or application to a person as an aid in diagnosis, or in evaluating the degree of susceptibility or immunity possessed by a person, and including also any other use for purposes of diagnosis if the diagnostic substance so used is prepared from or with the aid of a biological product."

These definitions identify and/or accommodate all the early biological preparations and several subsequent generations of biological products. Therefore, the regulatory classifications of blood products (e.g., whole blood, plasma, packed cells, and fractionated plasma), vaccines (e.g., live or killed bacterial and viral vaccines), toxins, antitoxins, and undefined allergenic extracts (e.g., immunizing agents, antisera, and allergens) are rarely, if ever, in question.

Many of the products made using biotechnology—products that one might assume would be biologics—often do not fall as easily within the boundaries of these definitions. Although made through recombinant techniques, both recombinant human insulin and recombinant human growth hormone are regulated solely under the FD&C Act and not the PHS Act, for instance. Since both substances are classified as hormones, they clearly fall under the domain of the FD&C Act, which specifically addresses hormone preparations. The classifications of the two products are further supported by the exclusion of hormones and amino acids from blood or plasma in the definition of "analogous product."

By applying these definitions to currently approved products made through newer technologies, one can gain a better understanding of the definitions and how they were applied. For example, alpha interferon, one of the first biotechnology products to be approved, was regulated as a biologic. Originally, alpha interferon was produced by inducing blood cells with sendai virus, causing the cells to secrete interferon. Later, researchers produced recombinant interferon. Naturally produced interferon that was isolated from the blood cells was used to make molecular probes that were, in turn, used to identify the gene responsible for the protein. After being isolated and characterized, the gene was inserted into the appropriate host to produce the recombinant gene product.

By studying the definitions, it becomes apparent that multiple interpretations of "analogous product" may apply to both recombinant and natural interferon. The original substances were generated by inducing blood cells with a virus. Thus, interferon could be considered analogous to a virus by definition. Since the source of the active substance was blood cells and the interferon would be released in the blood, it could also be considered analogous to a therapeutic serum. In addition, interferon could also be considered analogous to an "antitoxin," since it acts through the immune process.

Clearly, there is a strong rationale for classifying interferon as a biologic. For other products, the classification is not so clear.

For the complex and diverse products now produced through genetic engineering, molecular biology techniques, monoclonal antibody technology, and fusion technology, a careful assessment in view of the biologics' definitions is required to determine how, and under what standards, a product will be regulated.

14

General Comparison of Drug and Biologic Characteristics

Perhaps no two FDA-regulated products share more similarities than drugs and biologics. In a general sense, both are therapeutic entities used to cure or palliate medical conditions. Each is administered prophylactically or in response to a disease state, and is generally given orally or parenterally.

These and other similarities can make the process of classifying certain products as drugs or biologics rather difficult (see discussion below). Fortunately, however, several basic differences exist that, in most cases, make the product classification process fairly straightforward.

Conventional drugs are usually synthetic, organic compounds having defined structures and physical and chemical characteristics. Typically produced through chemical synthesis, these substances are usually micromolecules having molecular weights of less than 500 kilodaltons (kd). Researchers assess the actions of the compounds in standard pharmacological screens, and can generally assume the products' activities based on the similarity of overall chemical structures to those of other compounds. In general, the substances are synthesized and screened for a particular activity intended to treat a specific disease state. Conventional synthetic chemical drug substances are generally very stable, and not extremely sensitive to heat. These compounds can be studied in short- and long-term toxicology studies to assess safety.

In contrast, a biologic is usually a protein- or carbohydrate-based product, be it a viral or bacterial entity for use as a vaccine, an isolated therapeutic protein, or a blood or blood component. In all cases, the substances are either composed of, or are extracted from, a living organism. These substances are macromolecular by nature, and usually have a molecular weight greater than 500 kd. Since these are generally very large molecules, there is an increased complexity associated with determining physicochemical characteristics, such as tertiary structure, location, and extent and type of glycosylation. Therefore, they tend to be less well-defined (although analytical methodology and protein and carbohydrate chemistries have made significant advances in the last five to ten years).

Compared to conventional synthetic drugs, biologics tend to be rather labile, and are usually very heat- and shear-sensitive. The activity and utility of biological compounds are generally not known—one reason they are often considered therapeutics in search of a disease. Given their nature, the substances tend to be immunogenic, and cannot be assessed in long-term toxicology studies without using novel approaches.

Although advances in analytical technologies have improved biological product characterization, differences in drug and biologic manufacturing remain significant. For instance, a growing trend in the production of new biologics is the use of molecular techniques that typically involve the use of prokaryotic or eukaryotic microorganisms. Changes in these organisms can significantly affect the biological substance. And, as stated above, the quality, safety, and potency of this class of products are extremely sensitive to changes in manufacturing conditions.

Product Classification and the FDA's Intercenter Agreements

While traditional statutory and regulatory definitions resolve the majority of classification issues, technology has, in some cases, moved beyond the definitions. Biotechnology companies are creating a new generation of products not easily accommodated within established definitional boundaries. Meanwhile, other companies are busily combining different products, seemingly subjecting the products to multiple regulatory definitions and classifications.

In some cases, the scientific and medical challenges presented by emerging products has forced the FDA to employ innovative review approaches, such as calling on two centers to review a product's clinical data. This, in turn, has added to uncertainties regarding product classification issues—most importantly, how specific products would be regulated and reviewed.

In 1992, the FDA attempted to address this uncertainty through a series of three intercenter agreements (see Chapter 14). Signed by CBER, the Center for Drug Evaluation and Research (CDER), and the Center for Devices and Radiological Health (CDRH), the agreements specify how certain combination and difficult-to-classify products will be regulated. In addition, the agreements provide a framework for intercenter product reviews.

The intercenter agreement between CDER and CBER states that the following biological products require licensure and fall under CBER's authority:

- vaccines, regardless of the method of manufacture (vaccines were defined as agents administered for the purpose of eliciting an antigen-specific cellular or humoral response);

- *in vivo* diagnostic allergenic products and allergens intended for use as "hyposensitization agents";

- human blood or human blood-derived products, including placental blood-derived products, animal-derived procoagulant products and animal- or cell culture-derived hemoglobin-based products intended to act as red blood cell substitutes;

- immunoglobulin products;

- products composed of or intended to contain intact cells or intact microorganisms;

- proteins, peptides, or carbohydrate products produced by cell culture, excluding antibiotics, hormones, and products previously derived from human or animal tissue regulated as drugs;

- protein products made in animal body fluid by genetic alteration of the transgenic animals; and

- animal venoms or constituents of venoms.

Other classes of products identified as CBER-regulated products include:

- synthetically produced allergenic products intended to specifically alter the immune response to a specific antigen or allergen; and

- certain drugs used in conjunction with blood banking or transfusion.

The intercenter agreements provide the clearest account of products subject to the provisions of the PHS Act. Sponsors should use the agreements in conjunction with regulatory definitions to determine how and by which FDA center(s) a product will be regulated.

The Importance of Product Classification: Impact on Development Strategy

The formulation of the regulatory strategy is one of the most important elements of product development. This strategy comprises the overall development plan that specifies an approach to pharmacological, toxicological, and clinical testing, and product characterization and manufacture.

Obviously, this strategy is based largely on the nature of the product under development—whether it is a drug or biologic, for instance. The product classification is critically important because of fundamental differences in the regulatory requirements for drugs and biologics.

Drug approval in the United States requires the submission of a new drug application (NDA), which primarily focuses on chemical, pharmacological, and toxicological characterization of the product, and the demonstration of safety and effectiveness in humans. FDA regulations clearly specify that more than one adequate and well-controlled clinical trial is required to demonstrate safety and effectiveness in humans.

Toxicological characterization usually involves long-term, repeat-dose toxicology, reproductive, mutagenicity, and carcinogenicity studies. For conventional drugs, there are no restrictions on the use of multiparty manufacturing arrangements and contract manufacturers in the drug substance or drug product manufacturing process. Except for antibiotics, product certification—that is, individual batch release by the FDA—is not required.

In contrast, biologics licensure requires the simultaneous submission of a product license application (PLA) and an establishment license application (ELA). The FDA licenses the product, the process, the facilities where the product is manufactured, and the manufacturer.

The PLA focuses on the manufacturing process, as well as pharmacological, toxicological, and product characterization and the demonstration of safety and effectiveness in humans. In contrast, the ELA focuses on the features of the buildings, facilities, and equipment designed to prevent contamination of the product, and on other work performed at the manufacturing facility. Special emphasis is placed on equipment and methods validation to ensure the quality of the product and the prevention of product contamination.

Unlike FDA standards for conventional drugs, those for biologics require only one adequate and well-controlled human clinical trial to demonstrate safety and effectiveness. For biologics (proteins), the long-term toxicological testing is generally limited, due to the immunogenicity of the substance.

Another important difference is that licensing requirements and FDA policy limit the use of multiparty manufacturing arrangements and contract manufacturers in biologics manufacture. For multiparty manufacturing arrangements, special consideration is required, since each party involved in the manufacture of a biologic must hold both a PLA (see Chapter 8) and an ELA (see Chapter 9). While the FDA recently liberalized its policies regarding multiparty manufacturing arrangements for biologics, these policies remain more restrictive than those for drugs.

Finally, biologics are subject to product certification requirements. This process requires that manufacturers submit to the FDA samples from each batch of a product, samples which the agency tests prior to authorizing the release of the batches.

Obviously, these fundamental differences can influence the amount of testing required, the nature of manufacturing arrangements, and the overall development strategy for a product. Again, the single most important factor in determining these testing, manufacturing, and strategy issues is a product's classification.

To help determine the classification of specific products, manufacturers should assess the following characteristics of their products early in regulatory strategy development:

• the product's mechanism of pharmacological action;

• the product's original source (i.e., the source of the active substance);

• the product's characteristics and composition; and

• the product's method of manufacture.

Manufacturers should assess each of these parameters in light of the definition of a biologic provided by 21 CFR. 600.3 to determine how, and if, the definition applies to the product. The manufacturer should also obtain information about similar FDA-approved products to determine how these products were classified and reviewed. Further, the manufacturer should consult the FDA's intercenter agreements to determine whether these documents provide relevant criteria or guidance regarding the product.

If a product's classification remains unclear, the company should obtain FDA input (see Chapter 14). This is extremely important, given that interpreting product definitions is an inexact science.

References

(1) Biologics Control Act of 1902 (1902).

(2) Food, Drug and Cosmetic Act, as Amended (1938).

(3) Public Health Service Act (1944).

(4) 21 C.F.R. Parts 50-680, 1992.

(5) Buday, P.V., *Fundamentals of United States Biological Regulations*, 3 Regulatory Affairs 223-240 (1991).

(6) Grabenstein, J.D., *Don't Say "Biologics" If You Mean Immunologics*, 45 American Journal of Hospital Pharmacy 1941-42 (1988).

(7) Hecht, A., *Making Sure Biologics Are Safe*, FDA Consumer 21-26 (July-August 1977).

(8) Kondratas, R.A., *Biologics Control Act of 1902*, Federal Food and Drug Control 8-27.

(9) Kondratas, R.A., *Dealth Helped Write The Biologics Law*, FDA Consumer 23-25 (April 1982).

(10) Pitman, M., *The Regulation of Biologic Products, 1902-1972*, Ed.: Greenwald, H.R., Harden, V.A., U.S. Department of Health and Human Services, Public Health Service, National Institutes of Health, National Institute of Allergy and Infectious Diseases 61-70 (October 1987).

(11) Mathieu, M., New Drug Development: A Regulatory Overview (2nd Ed. 1990).

(12) Meyers, H.M., *Biologics and FDA,* FDA Consumer 12-17 (April 1973).

(13) Scott, S.R., *Joint Manufacturing of Biologics*, 5 U.S. Regulatory Reporter 3-5 (1988).

(14) A Brief Legislative History of the Food, Drug and Cosmetic Act, (Washington: G.P.O., January 1974).

Chapter 2

Preclinical Safety Assessment of Biological Products

by Joy A. Cavagnaro, Ph.D.
Office of Therapeutics Research and Review
Center for Biologics Evaluation and Research
Food and Drug Administration

Product identity, purity, potency, effectiveness, and safety are among the regulatory concerns applicable to all biological products. The Public Health Service Act defines biological product safety as "the relative freedom from harmful effects to persons affected, directly or indirectly, by a product when prudently administered, taking into consideration the character of the product in relation to the condition of the recipient at the time. Proof of safety shall consist of adequate tests by methods reasonably applicable to show that the biological product is safe under the prescribed conditions of use." Thus, the property of safety is relative.

As a biologic advances through the various phases of product discovery and development to approval and licensure, several issues are considered in assessing the biologic's safety. These issues, which are integral components of the regulatory decision-making process, include not only basic scientific issues but manufacturing, preclinical, medical, ethical, social, legal, and political concerns as well. Rarely are the issues independent.

As one element of a multidisciplinary research and product development program, preclinical safety evaluation studies are designed to provide a sufficient understanding of a product's pharmacology (including pharmacodynamic and pharmacokinetic profiles) and toxicological potential prior to the initiation of studies in humans. In addition to providing insights into a product's biology and safety, preclinical animal testing helps scientists to determine an optimal initial formulation, select an initial safe starting dose, develop a safe dose escalation scheme, identify potential target organ(s) of toxicity, recommend appropriate types of clinical monitoring procedures, and identify potential at-risk populations. Preclinical studies also may provide important information for the selection of a clinical indication. Following early clinical investigations or Phase 1 studies, subsequent preclinical studies are designed to facilitate the safe clinical development of the product by supporting various clinical decisions.

The number and nature of studies relevant in assessing biologics' safety and activity often differ between biologics of different classes and between biologics with different indications. In many cases, these studies also differ even for products within the same class.

The availability of alternative therapies and adequate technologies to assess risk(s) is also considered, as is the prevailing consensus regarding what constitutes acceptable risk. Despite the implementation of optimal preclinical testing strategies to assess safety and the use of rigorous product surveillance programs, biological products may cause unanticipated adverse clinical effects in humans. The ultimate goal of preclinical studies is to provide data that scientists can use to better predict potential adverse effects and to help researchers to design clinical studies that will minimize their occurrence.

Classification of Biological Products

A biological product is defined as any virus, therapeutic serum, toxin, antitoxin, or analogous product applicable to the prevention, treatment, or cure of diseases or injuries of man (21 CFR, Subchapter F, Subpart A, 600.3(h)). Since 1987, the FDA's Center for Biologics Evaluation and Research (CBER) has regulated the testing and marketing of biologics. In 1993, CBER was reorganized into three research and review offices, each of which has responsibility for a different product category: blood products, vaccines, and therapeutics (see Exhibit 1 below).

Historically, most biologics have been blood or vaccine products. The majority of these traditional biological products were used as replacement therapies or in limited doses prophylactically. As such, conventional toxicology tests performed to evaluate drug products were not routinely required.

The general safety test in mice and guinea pigs was, and still is, required for biological products to detect extraneous toxic contaminants. Additional safety testing required for specific bacterial products, viral vaccines, and human blood and blood products includes intracerebral inoculation of suckling mice, the neurovirulence safety test in monkeys, and the Schick test in guinea pigs. These additional preclinical safety tests were recommended based on specific product-related safety concerns.

Exhibit 1

Examples of the Diversity of Biological Products

• Office of Blood Research and Review

Blood (e.g., collection and processing)

Blood components (e.g., whole blood and cellular components)

Blood fractionation products (e.g., immunoglobulin, albumin, hemoglobin, clotting factors)

Diagnostic test kits (e.g., hepatitis, HIV)

• Office of Vaccines Research and Review

Antitoxins, antivenins, enzymes, and venoms

Vaccines (bacterial, viral, parasitic; prophylaxis and therapy)

Adjuvants

Allergenic products

• Office of Therapeutics Research and Review

Regulatory factors (e.g., cytokines, growth factors, neuroactive peptides)

Thrombolytics and fibrinolytics

Monoclonal antibodies (diagnostic, preventive, and therapeutic)

Novel technologies (e.g., cellular and gene therapies)

The types of biological products and their intended uses have expanded rapidly over the years (Exhibit 1 above depicts the diversity of biological products). They include naturally derived products, synthetic products, and products developed through biotechnology. While the quality control of biological products (i.e., analytical procedures, molecular biology, purification, characterization, lot reproducibility, stability, and identification and removal of adventitious agents and contaminants) is becoming less problematic, the products are becoming more sophisticated and complex. With advances in biotechnology, the development of novel adjuvants and delivery systems, and an anticipated rise in the use of combination products, there has been a more systematic application of pharmacological and toxicological studies in the preclinical evaluation of biologics.

Approaches to Preclinical Safety Evaluation

Historically, the philosophy of biologics regulation has been based on the use of available scientific data, theoretical and practical considerations, and flexibility in the application of requirements and standards. Due to the pace of scientific and technological advances, policy development has been, by necessity, a dynamic process. Therefore, approaches to biological product safety evaluation have evolved through the application of scientific insight, historical and anecdotal experiences, and common sense.

It is probably safe to state that there is no consensus on what constitutes an adequate preclinical safety program for biological products. This is the case primarily because the FDA has not published formal testing requirements or guidelines for pharmacology or toxicology studies applicable to all biological products. In contrast, several such guidelines have been available for drug products. These guidelines have provided both regulatory and industry scientists with a point of reference for determining what is likely to be acceptable, and may have promoted consistency in testing within and across drug product classes. In some cases, these guidelines have been applied to biological products.

Over the years, however, many preclinical safety studies probably have been performed just to "fulfill the regulatory requirements." It is often easier to perform the requisite list of studies than to justify a departure from the requirements, especially in light of the historical data bases and various precedents set over the years. Whatever was gained from the ability to pre-

26

dict the numbers and types of studies necessary for approval was likely lost in discouraging flexibility in the design and scientific defense of studies that would better predict potential adverse effects in humans.

Preclinical assessments involving conventional and creative approaches are equally appropriate. Without the benefit or burden of historical data bases or precedents, however, CBER has approached preclinical safety evaluation routinely on a case-by-case basis (see Exhibit 2). The case-by-case approach relies on a mutual understanding of general principles for designing studies rather than on specific agreements on standard protocols. As such, the approach does not foster in sponsors concrete expectations about the testing necessary in specific situations.

Exhibit 2

Approaches to Preclinical Safety Evaluation

- Traditional (conventional)
  - "guideline-driven"

- Case-by-Case (creative)
  - rational, science-based

- "Sufficient to slip by the regulators"
  - irrational, non-science-based

While the case-by-case approach is more demanding initially, it need not be more resource intensive. However, regulatory scientists and sponsors must have a common understanding of the potential safety concerns and a basic knowledge of the specific product. Acknowledging that the timing of preclinical studies is critical in a biological product's development, an alternative approach to preclinical safety assessment is presented in Figure 3 (see below).

Exhibit 3

The "What If" Approach

Fill in the Blank

Q. Sponsor

"If we don't have _____ in time for _____,
would it be OK if we just _____?"

A. Regulator

"Sure it would be OK if _____ but what if
_____?"

Controls for assuring the safety of biological products have been established through a variety of mechanisms, including specific types of regulatory documents (see Exhibit 4 below). To offer an alternative to formal regulations and guidelines, CBER has developed points-to-consider (PTC) documents to provide guidance in specific areas and to promote uniformity in regulatory reviews. The development of a PTC document generally parallels the development of the first product in a specific product class, and typically follows both CBER's review of a sponsor's initial scientific data and a discussion of key issues. PTC documents are designed to alert sponsors to areas of scientific concern in the product development process and to encourage the exchange of information. Although they are useful in strategic planning by industry, PTC documents do not have the force of law or regulations, and do not comprise a set of instructions that ensure product approval when followed (*Editor's note: for a listing of PTC documents and guidelines that provide information on preclinical issues, see the documents marked with asterisks in Appendix 1*).

Exhibit 4

Types of Regulatory Documents

Regulations:

Specifications for biological products are included in the requirements applicable to drugs (21 CFR, Part 312) and the general provisions and additional standards for biological products (21 CFR, Subchapter F, Parts 600-680).

Guidelines:

Documents that do not provide regulatory requirements but provide specific methods that may be used to meet requirements. If relevant procedures are followed, they generally will be considered to be approvable.

Points-To-Consider (PTC) Documents:

A mechanism to communicate principles under development (e.g., to improve quality, promote consistency). PTC documents are not designed to represent a set of instructions or guidelines that, if followed, will ensure that a product is approved.

Purpose of Preclinical Safety Evaluation

A sponsor's primary goal in conducting preclinical studies is to obtain data necessary to initiate clinical trials. These data provide the scientific basis for the development of clinical monitoring parameters and the rational selection of an initial safe starting dose, a dose escalation scheme, a duration of use, a route of administration, and potential target organs for toxicity.

Therefore, preclinical studies should be designed to answer specific questions. These answers should provide an understanding of the dose/activity relationship, the relationship of route and scheduling to activity/toxicity, the dose/toxicity relationship, and the risks for toxicity. Often, additional studies are designed to help discern a product's mechanism of action, to facilitate future clinical development (i.e., to help ensure that clinical trials are not

29

needlessly interrupted), and to satisfy liability and/or labeling issues (i.e., to establish or to promote a marketing advantage in some cases).

Over the years, misconceptions have arisen with regard to performance criteria for preclinical studies. These misconceptions have likely adversely affected the studies' potential for being predictive of beneficial or adverse effects, thus contributing to skepticism over the purpose of these studies and/or the relevance of these studies to clinical development programs.

Examples of the "myths of preclinical assessment programs" include the following: (1) the perception that more data are always more useful (i.e., more species, more routes, more schedules, longer durations); (2) the idea that unexplained observations are better ignored; (3) the view that a lack of significant adverse findings in animals represents an assurance of safety; and (4) the belief that, while animal models of disease (spontaneous or genetically induced) may provide insights into a biologic's pathobiology or early pathogenic events, toxicological findings will be too difficult to interpret and assess.

In providing general guidance for testing requirements, it is difficult to state, in absolute terms, that a study or set of studies either will always be or will never be required. In all likelihood, there will be exceptions. It is also difficult at times to decide whether certain information is necessary or is simply "nice to know." Resource limitations in government and industry will ultimately influence many of these decisions. For several reasons, a certain amount of "risk sharing" may be necessary at certain times in making decisions about the adequacy of data. However, the default position need not be toward traditional testing paradigms and standardized testing protocols.

Components of Preclinical Safety Assessment Programs

Components of preclinical development programs include studies to assess the following: *in vitro* and *in vivo* pharmacologic activity; pharmacokinetic parameters, including absorption, distribution, metabolism, and excretion (ADME); and *in vitro* and *in vivo* toxicity. The route of administration, dose, frequency, and duration utilized in preclinical studies are based on the product's proposed indication(s). The adequacy of the studies will depend on the review and integration of all the available data.

The critical issue in optimizing product-specific preclinical programs is deciding whether the available technologies for evaluation are useful and appropriate for the product being evaluated. If not, the sponsor must carefully consider the implementation of new technologies, including their identification, development, and validation. It is encouraging to note that many new technologies used in developing new biological products are being applied to their evaluation as well, even if only out of necessity. The advances in *in vitro* and *in vivo* preclinical tests that are expected as a result of studies designed to assess these products will complement ongoing efforts in other areas of toxicology.

Practical Issues and Common Questions

Due to the lack of formalized testing guidelines, there has been a natural tendency to expect that regulatory scientists know all the potential issues that research programs must address. Since these scientists know precisely what questions to ask, the logic goes, they can then assist in the design of appropriate studies. Although surprising to some, this is not always the case, especially since requests are often made without the benefit of any prior agency knowledge of the product, including chemistry and manufacturing information, data from preliminary *in vitro* or *in vivo* studies, and the proposed clinical population that the studies are expected to support.

There are several common, non-product-specific questions that sponsors ask regarding preclinical requirements for biologics (see Exhibit 5 below for examples). Some of these questions are asked for product-planning purposes to project test material requirements or to set critical time-line milestones rather than to obtain relevant scientific views about the product.

Ironically, since the guidance offered is data-driven and science-based, the scientists developing the product are probably in a better position to identify questions and issues that must be addressed. Ideally, a dialogue between FDA and industry scientists will take place early in product development. This mutual exchange of ideas takes advantage of expertise on the specific product, expertise on similar products derived from independent research, and possibly general knowledge derived from the evaluation of a collective data base of many products. Such discussions can lead to the design of optimal preclinical safety programs. Interpretations of the findings must then be considered in the context of the overall product development plan. The process is iterative. Knowledge gained from each phase of development is reviewed critically in the context of current and past experiences.

Exhibit 5

Frequently Asked Questions On Preclinical Safety Testing Requirements

Can we use a preliminary formulation for our preclinical studies? What is the test material in research grade? What if it is GLP grade? What types of bridging studies will we need when GMP clinical material is available?

What species should we use? How many species?

How do we select a relevant species?

Can we design preclinical studies with autologous proteins or cells?

How long should the studies be? How frequent should the doses be? What route(s) should we use? How high a dose in relationship to the proposed human clinical dose should be used? In terms of dose fractionation or multiple dose protocols, should studies mimic exactly the protocol proposed for use in humans?

How do we validate a novel/alternative test system?

Unfortunately, there is a paucity of published literature on the results of preclinical toxicology assessment programs for drugs and biologics. Much of the most interesting toxicological data may even be lost when sponsors decide against pursuing further product development. When data are made available, individual studies are selected and made available only after reaching some critical phase in development (e.g., after completion of Phase 2 clinical studies). Often times, the published data are not available until after the product is approved. This practice of withholding scientific data is clearly not optimal when regulatory guidance is sought early in product development, particularly in novel product development (including new molecular entities).

As industry and regulatory agencies become more comfortable with designing and reviewing "customized" preclinical strategies, the case-by-case approach may ultimately become the default approach. Those scientists expert in designing preclinical studies will then be able to focus on describing the scientific rationale for their respective preclinical safety assessment program designs rather than concerning themselves with recommending the "right" studies to satisfy the regulatory requirements.

In some cases, the use of standard protocols (e.g., reproductive toxicology, genotoxicity, and carcinogenicity studies) may be appropriate for assessing the safety of biological products. These protocols, however, are more significant in assuring "internationally acceptable data bases" that avoid unnecessary duplication of studies. When there are available data to suggest that modifications to the preferred protocol(s) are warranted, protocol changes will also be acceptable provided that there is sufficient scientific justification.

Limitations Inherent In Preclinical Safety Assessment
Despite efforts to maximize the predictive value of preclinical studies, there will always be inherent limitations. These may include: (1) a general lack of understanding of the fundamental biochemical and physiological mechanisms that could help to better define appropriate parameters for measurement in preclinical studies; (2) the absence of the target site or receptor in the test species; (3) significant differences in metabolic profiles across test species; and (4) difficulty in achieving sustained concentrations of the product at the target site or receptor(s). New considerations in the safety evaluation of biological products include immunogenicity, viral DNA contaminants, and species-specificity. When relevant animal models of disease or relevant species are available, the data are generally more useful. Interestingly, in the latter case, the animals are "normal." Therefore, they must have predictive value not only across species barriers but across physiological states.

Preclinical scientists recognize that humans are better predictors of a product's clinical safety and effectiveness. Therefore, the clinical studies are designed to predict, to the degree possible, clinical effectiveness and untoward toxicities. It is worth noting that, while preclinical studies are limited in scope compared to clinical studies (i.e., they are generally completed prior to product approval), clinical science has the lifetime of the product to update its predictions.

Points to Consider in Preclinical Safety Assessment of Biologics

Due to their diversity and complexity, biological products are evaluated on a case-by-case basis (see Exhibit 6 for a general outline for designing preclinical safety assessment programs). As sufficient data on either specific products or classes of products become available, the data are evaluated and specific recommendations are provided in the relevant PTC documents (e.g., the PTC documents on monoclonal antibodies, hemoglobin-based oxygen carriers, and cellular and gene therapies).

Exhibit 6

Points to Consider in the Preclinical Safety Assessment of Biologics

- Rationale
  - *In vitro* or *in vivo* studies
  - Potency assays
  - Receptor characteristics (across species)
  - Physiological modeling
  - Scientific literature
  - Scientific speculation

- Indication
  - Replacement therapy (long-term)
  - Non-pharmacodynamic treatment (prophylactic or diagnostic)
  - Pharmacodynamic treatment (short-term or long-term)

- Pharmacological Activity (pharmacodynamics)
  - Primary endpoints
  - Secondary endpoints

- *In vivo* Model Selection
  - Species-specific effects
  - Effects independent of species
  - Animal model of disease

- Pharmacokinetics and ADME: Correlation with Pharmacodynamics
  - Low dose
  - High dose

- General Toxicity
  - Single dose (acute)
  - Repeated dose (subacute or subchronic)

- Specific Toxicity (may include one or more of the following studies)
  - Local irritation (local reactogenicity)
  - Antigenicity
  - Chronic toxicity
  - Reproduction toxicity, including teratogenic potential
  - Mutagenicity
  - Tumorigenicity
  - Carcinogenicity
  - Other toxicity concerns (e.g., neurotoxicity, immunotoxicity, etc.)

The predictive value of preclinical studies is enhanced when sponsors use the product—or a closely representative product—that will be used in clinical trials. It is openly acknowledged that because biological products are complex, many changes will occur over the product's development. Since most changes are designed to improve the product, they are likely to be acceptable. Appropriate studies should be designed to assess the impact of these changes on product activity and safety throughout the course of product development.

While effectiveness need not be proven in early development, support for rationale is often provided and used to justify the potential risk to patients. *In vitro* studies may be sufficient. Potency assays that are required for biological products often serve to support the rationale. Better understanding of receptor characteristics helps in determining the most relevant species.

Since the interpretation of preclinical data is based largely on the proposed clinical application of the product, the data are most useful in predicting untoward human effects when the clinical application is also considered in the preclinical study design. Therefore, it is essential that, in selecting an initial indication, sponsors realize that the actual clinical application may not be determined, in some cases, until data from subsequent animal studies or even Phase 2 studies have been evaluated.

In general, *in vivo* animal studies should be designed to include sufficient numbers of animals per group to permit a valid estimation of an effect's (pharmacologic or toxicologic) incidence and frequency. Doses should be selected to provide data that will describe a dose-response. In addition to using comparable formulations, routes of administration, and schedules, comparable durations of exposure are also recommended.

The number of species necessary in preclinical testing programs varies. However, there is no specific requirement for the routine use of two species (e.g., one rodent and one non-rodent) in toxicology studies of biological products.

In each stage of product development, it is important to determine exposure by measuring pharmacokinetic (including ADME) or pharmacodynamic endpoints. This includes the following, for example: (1) measurements of the biologic in plasma or target organs; (2) the distribution and persistence of cells for cellular therapies; (3) measurements of viral shedding and recovery of certain vaccines; (4) localization of targeted novel delivery systems; and (5) tissue tropism, including germline tissue, of vectors used in gene therapies.

Such studies provide important information for a better interpretation of the toxicity observed in animals, and aid in the selection of not only the proposed initial human dose but of the dose escalation scheme and the frequency of dosing in the clinical trial(s). Further, once such exposure data are available in humans, the data can be used to better correlate the human and animal findings. Toxicity studies should be performed in the same species used to assess exposure. Often times and especially when non-rodents are used, exposure and toxicity are measured in the same study.

Toxicity studies should be designed not only to identify a safe dose, but also a toxic dose(s) to anticipate the product's safety and to better define the therapeutic index in humans. Specific product considerations that may complicate the process of defining a toxic dose may include limits based on formulation, lack of significant systemic absorption, or the amount of the product. The lack of significant toxicity in animals does not necessarily mean that the product is safe. The margin of safety for the initial starting dose, however, will likely be adequate.

Historically, the goal of acute toxicity studies was to define a lethal dose range following a single administration or the administration of a few closely spaced doses. More recently, these studies have been designed to evaluate a high dose that causes significant toxicity but not necessarily lethality. If deaths occur, rarely are such studies expected to provide sufficient information to determine the cause of death.

Studies are often one to two weeks in duration and routinely include body weight determinations, clinical observations, and gross necropsy findings. Additional antemortem studies may be performed as appropriate, especially in large animals (e.g., observation of local reactogenicity, pharmacokinetic evaluations, hematological and/or clinical biochemistry measurements). Histological evaluations may also be performed.

The duration of repeat-dose studies should be at least as long as the proposed clinical study. These studies are designed to establish a dose-response, define target organ(s) of toxicity, and determine whether observed toxicities are reversible. Evaluation parameters should include not only those routinely performed in the acute studies, but those performed in the additional studies as well. Special tests such as ophthalmoscopic, EKG, body temperature, and blood pressure monitoring are often included. Depending on the study duration, sampling at multiple time points may be indicated to better characterize the kinetics of response. As mentioned, a group of animals will be examined at term and some may be reserved for a treatment-free or recovery period to evaluate the reversibility of any findings.

Specific toxicity studies may be necessary due to special characteristics of the product or the clinical indication. For example, adjuvanted vaccines are routinely evaluated for local reactogenicity, and cellular therapies are routinely screened for tumorigenic potential. Further research is needed to better assess the genotoxic and tumor promoter potential of growth factors. In addition, research is needed to better predict the sensitizing potential of biological products and to determine the relevance of serum antibody levels following repeat dosing in animals and humans.

While carcinogenicity studies have not been performed routinely for biological products, they may be appropriate for products proposed for chronic use. Reproductive toxicology studies will probably become more common, especially as more women of child-bearing potential participate in early clinical trials. In the past, such studies have not been conducted for many biologics.

However, reproductive toxicology studies have been performed with many of the recently approved therapeutics. Such studies also have been conducted in the development of AIDS vaccines intended for use in pregnant women.

The recent development of biologics to treat various nervous system diseases has involved additional, specific neurotoxicological studies on these products. However, despite the fact that most products regulated as biologics have an immune component or impact directly or indirectly on the immune system, standardized immunotoxicity tests that are potentially useful for screening large numbers of chemicals for their ability to adversely affect the immune system have not proven essential in assessing the safety of biological products.

Throughout the various phases of product development, additional preclinical safety studies may be necessary due to unexpected toxicity, significant changes in the manufacturing process or the final formulation, or changes in the clinical indications (see Exhibit 7 below). In some cases, the ideal assessment of the safety of novel biological therapies may require alternative approaches such as *in vitro* or *in vivo* organogenesis model systems, animal models of tolerance, animal models of disease, or transgenic animal models.

Exhibit 7

Additional Preclinical Safety Studies

- Unexpected toxicity
    - May be species-specific
    - Is related to route, dose, frequency
    - May be related to immunogenicity

- Significant changes in the final formulation
    - Altered pharmacokinetics
    - Safety not established
    - Change in effectiveness

- Change in clinical indication
    - May include change of route, dose, frequency
    - Intended clinical population (e.g, life-threatening (acute) to minor (chronic))
    - Combination therapies (synergistic or competitive effects)

Summary

Over the past decade, the scientific community has learned much about the predictive value of a variety of testing strategies in evaluating biological products. The learning curve has been steep. It has relied on the exchange of ideas between industry and regulatory scientists who are working toward the common goal of expediting the availability of safe and effective new products. The introduction of unique new therapies has been facilitated by an emphasis on cooperation between industry, academic, and regulatory scientists, and an adherence to sound scientific principles, common sense, and an approach based on flexibility.

During the next decade, it is likely that approaches to preclinical safety evaluation will continue to improve and to provide the critical information necessary to enable the expeditious development and approval of safe and effective new therapies. Careful design and the judicious use of animals should help provide for the early initiation of clinical studies and an uninterrupted clinical development.

The challenges facing regulatory agencies and industry sponsors over this next decade will include: (1) implementing appropriate programs that will better anticipate safety concerns, especially those inherent in novel technologies; (2) developing innovative *in vitro* and *in vivo* toxicity assays; and (3) designing appropriate preclinical testing strategies to better quantify safety concerns. To ensure that new biological products are both safe and effective and that they are made available without delay, regulatory agencies must actively participate in dialogues with industry and must maintain a regulatory environment that encourages innovation.

Industry must initiate interactions with the FDA earlier in the development process (i.e., during the research and discovery phases). In addition, industry must not only understand the regulatory review process but also must prepare product development plans that are realistic, flexible, and consistent with this process. For both regulatory agencies and industry, it is critical that the process of biological product development be well-managed.

The powers of preclinical safety studies to predict clinical effects are only as good as the design of the program. The same amount of careful consideration used in designing the clinical program should be afforded the preclinical program. DO GOOD SCIENCE.

This chapter reflects the author's assessment of the requirements for pre-clinical safety studies and is not intended to represent the official position of the FDA.

References

(1) Cavagnaro, J.A., "Misconceptions with Biotechnology" from the proceedings of the First International Conference on Harmonization, Brussels 1991. Edited by P.F. D'Arcy and D.W.G. Harron, The Queens University of Belfast, pp. 301-307 (1992).

(2) Cavagnaro, J.A., *Science-Based Approach to Preclinical Safety Evaluation of Biotechnology Products*, Pharmaceutical Engineering 32-33 (May/June 1992).

(3) Cavagnaro, J.A., *Regulatory Concerns in the Current Practice of Clinical Pathology*, 20 Toxicological Pathology 519-522 (1992).

(4) 21 CFR Parts 600-799.

(5) Goldenthal, E.I., *Current Views on Safety and Evaluation of Drugs*, FDA Papers 1-8 (May 1968).

(6) Goldenthal, K.L., Cavagnaro, J.A., Alving, C.R., and Vogel, F.R., *Safety Evaluation of Vaccine Adjuvants. NCVDG Meeting Working Group*, (submitted to AIDS Research and Human Retroviruses) (in press).

(7) Hayes, T.J. and Cavagnaro, J.A., *Progress and Challenges in the Preclinical Assessment of Cytokines*, 64265 Toxicology Letters 291-297 (1992).

(8) WHO Technical Report Series No. 563, *Guidelines for Evaluation of Drugs for Use in Man* (1975).

(9) Zbinden, G., *Biotechnology Products Intended for Human Use, Toxicological Targets and Research Strategies*, Preclinical Safety of Biotechnology Products Intended for Human Use 143-159, Alan R. Liss, Inc. (1987).

Chapter 3:

The FDA's Regulation of Preclinical Testing: Good Laboratory Practice (GLP)

by Mark Mathieu
PAREXEL International Corporation

Manufacturers of biologics and all other FDA-regulated products are given considerable freedom during the preclinical screening and testing of new products. Provided they comply with the U.S. Animal Welfare Act and other applicable animal welfare laws, nonclinical testing laboratories at biological companies and private contractors are not limited in the use of animals to screen and measure the activity of biologics.

When the sponsor begins to compile safety data for submission to the FDA, however, a set of standards called Good Laboratory Practice (GLP) applies. To ensure the quality and integrity of data derived from nonclinical testing, the FDA requires that nonclinical laboratory studies designed to provide safety data for an IND, PLA, or other regulatory submission compl with GLP standards. GLP regulations apply to product sponsor laboratories, private toxicology laboratories, academic and government laboratories, and all other facilities that conduct animal testing and perform related analyses, the results of which will be submitted to the FDA in support of a product's safety.

GLP regulations seek to ensure the quality and integrity of data by establishing basic standards for the conduct and reporting of nonclinical safety testing. Specifically, GLP regulations set standards in such areas as the organization, personnel, physical structure, maintenance, and operating procedures of nonclinical testing facilities.

GLP: A Short History

GLP regulations, which became effective on June 20, 1979, were the FDA's response to finding, in the mid-1970's, that some nonclinical studies submitted to support the safety of new drugs were not being conducted according to accepted standards. The FDA considered this a serious problem because data from these studies served as the basis for important regulatory decisions— specifically, whether clinical trials could be initiated or continued, and whether new drugs should be approved.

After establishing the initial GLP regulations in 1979, the FDA's confidence in the work of preclinical laboratory facilities increased markedly. As a result, on October 5, 1987, the agency published revised GLP regulations that sought to reduce regulatory and paperwork burdens facing laboratories conducting animal studies.

The FDA's 1987 GLP regulations brought changes in such areas as quality assurance, protocol preparation, test and control article characterization, and specimen and sample retention. In virtually all cases, the changes were intended to ease GLP requirements.

The Applicability of GLP

On one level, GLP applicability is fairly straightforward. According to FDA regulations, GLP applies to facilities conducting "nonclinical laboratory studies that support or are intended to support applications for research or marketing permits for products regulated by the Food and Drug Administration, including food and color additives, animal food additives, human and animal drugs, medical devices for human use, biological products, and electronic products."

What at times seems more difficult is identifying which nonclinical tests conducted within those facilities are subject to GLP requirements. The FDA states that GLP applies to "nonclinical laboratory studies," which the regulations define as "*in vivo* or *in vitro* experiments in which test articles are studied prospectively in test systems under laboratory conditions to determine their safety. The term does not include studies utilizing human subjects or clinical studies or field trials in animals. The term does not include basic exploratory studies carried out to determine whether a test article has any potential utility or to determine physical or chemical characteristics of a test article."

Specifically, GLP applies to all "definitive" nonclinical safety studies, including key acute, subacute, chronic, reproductive, and carcinogenicity studies. Preliminary pharmacological screenings and metabolism studies are exempt from GLP requirements, as are initial pilot studies such as dose-ranging, absorption, and excretion tests.

The use of an outside testing laboratory for nonclinical testing does not eliminate the need for GLP compliance. When using the services of a consulting laboratory, contractor, or grantee to perform an analysis or other service, the sponsor must notify the vendor "that the service is part of a nonclinical laboratory study that must be conducted in compliance with [GLP requirements]."

In some cases, sponsors may obtain exemptions from specific GLP provisions for special nonclinical laboratory studies. An FDA *Questions and Answers* document on GLPs states that "not all of the GLP provisions apply to all studies and, indeed, for some special studies certain of the GLP provisions may compromise proper science. For this reason, laboratories may petition the agency to exempt certain studies from some of the GLP provisions. The petition should contain sufficient facts to justify granting the exemption."

The Major Provisions of GLP

The core provisions of GLP set standards for the nonclinical laboratory's organization, physical structure and equipment, and operating procedures. For purposes of analysis, these standards may be grouped into seven general areas:

• organization and personnel;
• testing facility;

- testing facility operation;
- test and control article characterization;
- the protocol and conduct of the nonclinical laboratory study;
- records and reporting; and
- equipment design.

Organization and Personnel   GLP regulations regarding a nonclinical laboratory's organization and personnel address four areas: general personnel, testing facility management, study director, and quality assurance unit. Aside from general qualifications and responsibilities of personnel and management, however, this aspect of GLP focuses primarily on issues regarding the study director and quality assurance unit.

Study Director.   GLP stipulates that the management of the testing facility conducting a preclinical program designate a scientist or other professional to serve as the study director. This individual has overall responsibility for the "technical conduct of the study, as well as for the interpretation, analysis, documentation, and reporting of results and represents the single point of study control." The FDA does not require that the study director be technically competent in all areas of a study, however.

The study director and others involved in conducting and supervising animal experiments should possess the education, training, and experience necessary to perform their assigned functions. Current training, experience, and job description profiles must be maintained for each of these persons. These documented profiles must be stored by the facility and made available to the FDA if the agency has any questions about personnel qualifications.

Quality Assurance Unit.   GLP also requires that each testing facility have a quality assurance unit (QAU) of one or more persons directly responsible to the facility management. The QAU monitors each study to "assure management that the facilities, equipment, personnel, methods, practices, records, and controls" are consistent with GLP. To ensure that such evaluations are made objectively, QAU members may not be involved in any animal study that they monitor.

The GLP regulations specify seven major QAU responsibilities in the areas of record maintenance, study inspections, and reports to facility management. At the conclusion of a study, the QAU certifies that the study was conducted under GLP.

QAU inspectional policy revisions represented the most important changes that the 1987 GLP regulations brought to the quality assurance area. Current regulations require that the QAU inspect a nonclinical study at intervals the unit considers adequate to ensure the study's integrity. No longer must QAUs inspect studies at specific intervals. However, the FDA advises that each study, regardless of its length, be inspected in-process at least once, and that, across a series of studies, all phases be inspected to assure study integrity.

In the 1987 GLP regulations' preamble, the FDA also clarified its views on the composition and function of the QAU in the following ways:

- "FDA never has intended that the QAU necessarily has to be a separate entity or a permanently staffed 'unit'..."

- "FDA continues to believe that well-qualified and trained personnel are essential to quality assurance under the GLP regulations and that one of management's most important responsibilities in maintaining effective quality assurance is to provide an adequate number of such personnel."

- The FDA did not and does not require under the revised regulation that the QAU be composed of individuals whose only duties are in quality assurance. The "agency intends only that quality assurance activities be separated from study direction and conduct activities; that is, a trained and qualified person who works on one study can perform quality assurance duties on any study in which he or she is not involved."

Testing Facility   Obviously, the laboratory facility in which the testing program takes place is also a primary focus of GLP. Animal care and supply facilities, test substance handling areas, laboratory operation, specimen and data storage, administrative and personnel facilities, methods of dosage preparation, and test substance accountability all must meet detailed requirements. Records indicating compliance must be kept.

In general terms, a testing facility and its equipment must be of suitable size and construction to allow for the proper conduct of the nonclinical study. Animal care areas, for example, must provide for sufficient separation of species/test systems and individual projects, isolation of animals, protection from outside disturbances, and routine or specialized animal housing. Regulated environmental controls for air quality—temperature, humidity, and air changes—and sanitation are needed.

Operation of Testing Facility   Each laboratory must base its operations on standard operating procedures (SOPs). SOPs, which are in some respects extensions of preclinical protocols, are written study methods or directions that laboratory management believes are adequate to guarantee the quality and integrity of data obtained from animal tests. The description of research procedures provided by protocols and SOP documents makes possible the verification and reconstruction of studies.

The detailed written procedures specified in the SOPs must be maintained for all aspects of the study, including animal care, laboratory tests, data handling and storage, and equipment upkeep and calibration. Each laboratory area must have accessible laboratory manuals and SOPs relative to the laboratory procedures being performed. Determining the degree to which SOPs are observed is another of the QAU's duties.

Any deviations from established SOPs must be authorized by the study director and noted in the raw data. On the other hand, major deviations must first be approved by the laboratory's management.

Test and Control Article Characterization   Under the 1987 GLP revision, testing facilities are not required to characterize test and control articles before toxicity studies begin. This allows companies to screen out many of the useless compounds before investing resources necessary to characterize them: "FDA has concluded that characterization of test and control articles need not be performed until initial toxicology studies with the test article show reasonable promise of the articles reaching the marketplace. In arriving at this conclusion, the agency considered that prior knowledge of the precise molecular structure is not vital to the conduct of a valid toxicology test. It is important, however, to know the strength, purity, and stability of a test or control article that is used in a nonclinical laboratory study." FDA officials point out that either the sponsor or the testing laboratory may handle the article characterization tasks.

The FDA also revised its requirements for stability testing. Previously, stability testing of test and control articles had to be conducted before study initiation or, if this was impossible, through periodic analyses of each batch. The 1987 regulations allow facilities to choose either route.

The Protocol and the Conduct of Nonclinical Laboratory Studies

A protocol, or testing plan, is a vital element in both clinical and preclinical studies. GLP states that a preclinical program must have a written protocol that "clearly indicates the objectives and all methods for the conduct of the study. "

Included in the 12-item protocol, which the sponsor must approve, should be descriptions of the experimental design and the purpose of the study as well as the type and frequency of tests, analyses, and measurements to be made. Changes made to the protocol during the course of the study call for an official protocol amendment signed by the study director. The study director's approval of protocol amendments assures the FDA of the data's integrity.

Although protocols and SOPs may seem similar, the two have different purposes. The protocol is specific to the study being conducted, while laboratory SOPs are standards used for all research projects at a given facility. For example, SOPs would provide "how to" instructions on a facility's routine procedures for obtaining animal blood samples, caring for animals, and using and maintaining equipment.

The protocol provides instructions that are study-specific. The protocol would detail, for a particular study, how often and from what animals blood samples are to be taken; what tests are to be conducted; and the number, species, sex, age, and weight of the animals to be tested. These factors would probably differ for each preclinical study, while the facility's SOPs would remain the same. In general terms, the protocol identifies what tests are necessary; SOPs instruct a facility's employees how these tests should be performed.

The 1987 GLP regulations allow facilities to use "umbrella protocols." By revising the definition of "nonclinical laboratory study," the FDA permits the conduct of multiple studies using the same test article under one comprehensive protocol. According to the preamble of the 1987 GLP final rule: "Under the revised definition...a single 'umbrella protocol' may be used for concurrent testing of more than one test article using a single, common procedure, e.g. mutagenicity testing, or for a battery of studies of one test article conducted in several test systems." Regarding the advantages, the regulation adds that the "agency recognizes that a longer, more complex protocol might be more difficult to manage than a simpler one; however, using an 'umbrella' protocol should be more efficient than using several closely-related protocols."

Records and Reporting   A final report must be prepared for each nonclinical laboratory study. Comprehensive reports typically include the summary, testing methods, results, and conclusions of a study, as well as all raw data on each of the test animals.

These final reports, as well as all raw data, documentation, protocols, and certain specimens generated during the toxicology study must be stored in an archive or repository to assure their safety and integrity for specific periods as designated in GLP regulations. Although these regulations state that two to five years is adequate, the FDA sometimes recommends that records be stored indefinitely and that specimens—slides, tissues, and blocks—be stored as long as they can be used to validate data.

In at least one respect, the 1987 GLP regulations eased record and sample retention requirements. Wet and mutagenicity test specimens must no longer be retained by nonclinical testing laboratories. According to the regulations: "All raw data, documentation, protocols, final reports, and specimens (except those specimens obtained from mutagenicity tests and wet specimens of blood, urine, feces, and biological fluids) generated as a result of a nonclinical laboratory study shall be retained."

Equipment Design   GLP regulations specify requirements for the design, maintenance, and calibration of equipment used in nonclinical tests. Automatic, mechanical, or electronic equipment used in the generation, measurement or assessment of data and equipment used for facility environmental control must be: (1) of appropriate design and adequate capacity to function according to the protocol; (2) suitably located for operation, inspection, cleaning, and maintenance; and (3) adequately tested, calibrated, and/or standardized.

SOPs are required to define, in sufficient detail, the methods, materials, and schedules to be used in the routine inspection, cleaning, maintenance, testing, calibration, or standardization of equipment. The facility must maintain written records of routine and certain nonroutine procedures involving laboratory equipment.

FDA GLP Inspections

To monitor compliance with GLP requirements, the FDA employs a program of on-site laboratory inspections and data audits. According to the FDA's Compliance Policy Guide 7348.808, the FDA conducts two basic types of GLP compliance inspections: surveillance inspections and directed inspections.

Representing the majority of GLP inspections, on-site surveillance inspections are periodic, routine evaluations of a laboratory's GLP compliance. Typically, these evaluations are based on studies in progress or recently completed studies. Routine inspections for monitoring a nonclinical laboratory's GLP compliance are scheduled "approximately every two years."

By policy, FDA investigators are instructed to contact facilities that will be the subject of a surveillance inspection no more than one week prior to the inspection. Under some circumstances, the FDA will agree to a short postponement of the inspection.

According to CPG 7348.808, the FDA might assign a directed inspection for any of the following reasons:

- as a follow-up to the disclosure, during a routine surveillance inspection, that a facility is violating GLP regulations or that there exist significant or major discrepancies between a study's final reports and original data and records;

- as a follow-up to the discovery of questionable data or materials which raise suspicions during a review of a study report submitted in support of an IND or PLA;

- as a response to the need to audit the data of an important or critical study submitted in support of an IND or PLA;

- as a response to the need to verify the audit performed by a sponsor or third party of the data and records of a study; or

- as a response to the need to review or inspect entities or studies brought to FDA's attention by other sources (i.e., news media or other operating firms/labs).

Directed inspections are conducted when necessary. Typically, these are unannounced inspections.

CBER and GLP Inspectional Activities

CBER does not employ a highly active inspectional program for monitoring GLP compliance. Since most facilities that conduct nonclinical biological studies perform drug studies as well, CBER reviews the routine inspections conducted by CDER. When it inspects a facility that conducts studies for both types of products, CDER routinely provides a copy of the inspection results to CBER.

On a limited basis, CBER conducts its own directed inspections. The center averages only one to two such inspections per year.

A principal reason for limited CBER interest in biologics GLP inspections is that typically there are no long-term toxicity studies conducted on biological products. Inspections of studies of limited duration are often not viewed as being cost effective.

The Disqualification of Nonclinical Studies   If FDA inspectors find GLP violations within a facility, agency officials review the violations and decide on a course of action. In cases involving severe compliance problems, the FDA may opt to disqualify data from an entire nonclinical study. According to federal regulations, the purposes of the agency's right to disqualify data are: "(1) To permit the exclusion from consideration completed studies that were conducted by a testing facility which has failed to comply with the requirements of the good laboratory practice regulations until it can be adequately demonstrated that such noncompliance did not occur during, or did not affect the validity or acceptability of data generated by, a particular study; and (2) to exclude from consideration all studies completed after the date of disqualification until the facility can satisfy the Commissioner that it will conduct studies in compliance with such regulations."

As implied by the second provision above, the FDA also can disqualify preclinical laboratories. For the FDA to do this, three conditions must be met. Specifically, the agency must find that: "(a) The testing facility failed to comply with one or more of the regulations...; (b) The noncompliance adversely affected the validity of the nonclinical laboratory studies; and (c) Other lesser regulatory actions (e.g., warnings or rejection of individual studies) have not been or will probably not be adequate to achieve compliance with the good laboratory practice regulations."

Chapter 4:
The Biological IND

During a new biologic's early preclinical development, the sponsor's primary goal is to determine if the product is reasonably safe, and if it exhibits activity that justifies commercial development. Should the product prove to be a viable candidate for further development, the sponsor then focuses on collecting the data and information necessary to establish that the product will not expose human subjects to unreasonable risks when used in limited, early-stage clinical studies.

Generally, this includes data and information in three broad areas:

Animal Pharmacology and Toxicology Studies. Preclinical data to permit an assessment as to whether the biological product is reasonably safe for initial testing in humans.

Manufacturing Information. Descriptions of the biological product's composition and source, and the methods used for its manufacture to allow an assessment as to whether the company can adequately produce and supply consistent batches of the product.

Clinical Protocols and Investigator Information. Detailed protocols for proposed clinical studies to permit an assessment as to whether the initial-phase trials will expose subjects to unnecessary risks. Also, information on the qualifications of clinical investigators—professionals (generally physicians) who oversee the administration of the experimental compound—to permit an assessment as to whether they are qualified to fulfill their clinical trial duties.

Before initiating clinical trials of a new biological product, the sponsor must compile this information in a document called an investigational new drug application (IND). In many ways, the IND is simply a research proposal through which the company obtains the FDA's permission to begin clinical testing.

The Role of the IND

The IND is not an application for marketing approval. Rather, it is a request for an exemption from the federal statute that prohibits an unapproved biologic from being shipped in interstate commerce. Current federal law requires that a drug or biologic be the subject of an approved PLA or NDA before it is transported or distributed across state lines. Because the sponsor will probably want to ship the investigational biologic to clinical investigators in other states, it must seek an exemption from that legal requirement. The IND is the means through which the sponsor obtains this exemption from the FDA.

It is worth noting that there are several types of INDs. This chapter focuses on "commercial INDs"—applications submitted mainly by manufacturers whose ultimate goal is to obtain marketing approval for a new product.

Several varieties of applications may be grouped within a second class of filings often called "noncommercial INDs." One type of noncommercial filing is the research IND, which is typically submitted by a physician or research institution and which generally proposes studies on previously studied products. A physician might file a research IND to propose studying an approved or unapproved product in a new patient population or for a new indication, although not necessarily with the motivation of supporting additional labeling for the product. While research INDs must fulfill the same submission requirements as commercial INDs, supporting data for significant portions of the IND may be submitted by cross-referencing other applications. With permission from the sponsor of a related IND, PLA, or master file, information on a product's manufacture and data regarding pharmacology, toxicology, and previous human experience may be incorporated into the research IND to support clinical trials. A letter from the reference IND, PLA, or master file sponsor to the research IND sponsor stating that the FDA may reference the files on behalf of the research IND sponsor will fulfill the submission requirement. Such letters should reference specific file numbers and specify the sections of the filings that may be referenced.

Other noncommercial INDs, such as "treatment INDs" and "emergency use INDs," allow sponsors to make available, and private physicians to acquire, experimental-stage biologics for use in desperately ill patients or in patients for whom no adequate therapies are otherwise available.

Product Jurisdiction and the IND

As early as possible in product development (and certainly before the submission of the IND), a sponsor must assess whether the investigational product is, in fact, a biologic (see Chapter 1 and Chapter 14). The product's regulatory status or classification—that is, whether it is a biologic, drug, etc.—will have fundamental implications on the preclinical and laboratory testing that must be conducted and analyzed in the IND. The product's regulatory classification will also determine which FDA center has jurisdiction over the IND.

A product's classification is based principally on the nature of the product itself. The biologic's source materials, physical characteristics, pharmacologic properties, and method of manufacture are the primary factors. Particularly for combination products (e.g., combinations of a biological product and a device), such determinations are not always straightforward.

The FDA has developed several forms of guidance to assist sponsors in making, or obtaining FDA decisions on, product classifications. For example, the agency has developed and published three intercenter agreements that outline the major product categories and provide basic guidance on combination products. The agreements apply to drugs, biologics, and medical devices and the three centers that regulate them: CDER, CBER, and CDRH (see Chapter 14).

When the intercenter agreements are not sufficient to help sponsors determine the FDA center with jurisdiction for their products, sponsors may contact the center(s) likely to be responsible for the review. In most cases, the jurisdiction can be resolved between the centers.

If the issue remains unresolved, sponsors may contact the FDA's "jurisdiction officer," otherwise known as the FDA chief mediator and ombudsman. The jurisdiction officer will review and act upon each request in writing within 15 days.

Although one center may be responsible for reviewing a product's application, other centers may be consulted for assistance during the review process. The "lead" center—the center with primary jurisdiction over a product—will make the final decisions on an application, however.

The Content and Format of the IND

The FDA regulations that identify and define IND submission requirements were published in 1987. These regulations were updated primarily to encourage the submission of better organized INDs that would be easier for the FDA to review.

The 1987 regulations address far more than IND format and submission requirements, however. As described elsewhere in this text, the regulations govern the FDA's review of INDs, the regulation of the various phases of clinical research, sponsor and clinical investigator responsibilities during a biologic's clinical development, and FDA-sponsor communication during clinical trials.

According to these regulations, the IND application should include the following information applicable to the biological product:

1. Cover Sheet (Form FDA-1571).
2. Table of Contents.
3. Introductory Statement.
4. General Investigational Plan.
5. Investigator's Brochure.
6. Clinical Protocols.
7. Chemistry, Manufacturing, and Control Information.
8. Pharmacology and Toxicology Information.
9. Previous Human Experience with the Investigational Product.
10. Additional Information.
11. Relevant Information.

Cover Sheet (Form FDA-1571)   The IND application form—referred to as Form FDA-1571 (see sample form below)—serves as the cover sheet for the entire application. By signing and completing, or having an authorized representative sign and complete, this form, the sponsor: (1) identifies itself, the investigational biologic, the product's proposed use, the clinical investigators and monitors, and the phase or phases of investigation covered by the application; (2) commits to following applicable regulations; and (3) identifies any responsibilities that have been transferred to a contract research organization. The form is available directly from the FDA.

Table of Contents   The ordering of contents outlined above facilitates the IND review. The table of contents should have adequate detail to allow FDA reviewers to locate important elements of the application easily and conveniently.

Introductory Statement   According to FDA regulations, the first part of this two-part section should provide "a brief introductory statement giving the name of the [product] and all active ingredients, the [product's] pharmacological class, the structural formula of the [product] (if known), the formulation of the dosage form(s) to be used, the route of administration, and the broad objectives and planned duration of the proposed clinical investigation(s)" [21 CFR 312.23(a)(3)(i)]. The sponsor must also summarize all previous use and marketing of the [product] in human populations, including any testing in the United States or abroad, and foreign marketing experience that might be relevant to the safety of the clinical investigations proposed in the IND. If the product's marketing or testing were discontinued for any reason related to safety or effectiveness, the sponsor must identify the country or countries where the withdrawal took place, and must describe the reason for the withdrawal.

General Investigational Plan   The general investigational plan must provide a brief description of the overall plan for investigating the biologic for the following year. This summary gives the reviewers a very brief overview of clinical studies to be conducted, and provides the necessary context for the reviewers to assess whether sufficient information to support future studies has been provided. The sponsor is free to deviate from the plan when necessary, provided that the sponsor fulfills relevant protocol and information amendment reporting requirements.

Biologics Development: A Regulatory Overview

<table>
<tr><td colspan="2">DEPARTMENT OF HEALTH AND HUMAN SERVICES
PUBLIC HEALTH SERVICE
FOOD AND DRUG ADMINISTRATION
INVESTIGATIONAL NEW DRUG APPLICATION (IND)
(TITLE 21, CODE OF FEDERAL REGULATIONS (CFR) PART 312)</td><td>Form Approved: OMB No. 0910-0014.
Expiration Date: December 31, 1992.
See OMB Statement on Reverse.

NOTE: No drug may be shipped or clinical investigation begun until an IND for that investigation is in effect (21 CFR 312.40).</td></tr>
</table>

1. NAME OF SPONSOR	2. DATE OF SUBMISSION
3. ADDRESS *(Number, Street, City, State and Zip Code)*	4. TELEPHONE NUMBER *(Include Area Code)*
5. NAME(S) OF DRUG *(Include all available Trade, Generic, Chemical, Code)*	6. IND NUMBER *(If previously assigned)*

7. INDICATION(S) *(Covered by this submission)*

8. PHASE(S) OF CLINICAL INVESTIGATION TO BE CONDUCTED. ☐ PHASE 1 ☐ PHASE 2 ☐ PHASE 3 ☐ OTHER_____
 (Specify)

9. LIST NUMBERS OF ALL INVESTIGATIONAL NEW DRUG APPLICATIONS *(21 CFR Part 312)*, NEW DRUG OR ANTIBIOTIC APPLICATIONS *(21 CFR Part 314)*, DRUG MASTER FILES *(21 CFR 314.420)*, AND PRODUCT LICENSE APPLICATIONS *(21 CFR Part 601)* REFERRED TO IN THIS APPLICATION.

10. *IND submissions should be consecutively numbered. The initial IND should be numbered "Serial Number: 000." The next submission (e.g., amendment, report, or correspondence) should be numbered "Serial Number: 001." Subsequent submissions should be numbered consecutively in the order in which they are submitted.*	SERIAL NUMBER: ___ ___ ___

11. THIS SUBMISSION CONTAINS THE FOLLOWING (Check all that apply)

☐ INITIAL INVESTIGATIONAL NEW DRUG APPLICATION (IND) ☐ RESPONSE TO CLINICAL HOLD

PROTOCOL AMENDMENT(S):	INFORMATION AMENDMENT(S):	IND SAFETY REPORT(S):
☐ NEW PROTOCOL	☐ CHEMISTRY/MICROBIOLOGY	☐ INITIAL WRITTEN REPORT
☐ CHANGE IN PROTOCOL	☐ PHARMACOLOGY/TOXICOLOGY	☐ FOLLOW-UP TO A WRITTEN REPORT
☐ NEW INVESTIGATOR	☐ CLINICAL	

☐ RESPONSE TO FDA REQUEST FOR INFORMATION ☐ ANNUAL REPORT ☐ GENERAL CORRESPONDENCE

☐ REQUEST FOR REINSTATEMENT OF IND THAT IS WITHDRAWN, INACTIVATED, TERMINATED OR DISCONTINUED ☐ OTHER_____
 (Specify)

CHECK ONLY IF APPLICABLE

JUSTIFICATION STATEMENT MUST BE SUBMITTED WITH APPLICATION FOR ANY CHECKED BELOW. REFER TO THE CITED CFR SECTION FOR FURTHER INFORMATION.

☐ TREATMENT IND 21 CFR 312.35(b) ☐ TREATMENT PROTOCOL 21 CFR 312.35(a) ☐ CHARGE REQUEST/NOTIFICATION 21 CFR 312.7(d)

FOR FDA USE ONLY

CDR/DBIND/OGD RECEIPT STAMP	DDR RECEIPT STAMP	IND NUMBER ASSIGNED
		DIVISION ASSIGNMENT

FORM FDA 1571 (6/92)   PREVIOUS EDITION IS OBSOLETE

58

12.

CONTENTS OF APPLICATION

This application contains the following items: *(check all that apply)*

☐ 1. Form FDA 1571 [21 CFR 312.23 (a) (1)]

☐ 2. Table of contents [21 CFR 312.23 (a) (2)]

☐ 3. Introductory statement [21 CFR 312.23 (a) (3)]

☐ 4. General investigational plan [21 CFR 312.23 (a) (3)]

☐ 5. Investigator's brochure [21 CFR 312.23 (a) (5)]

☐ 6. Protocol(s) [21 CFR 312.23 (a) (6)]

    ☐ a. Study protocol(s) [21 CFR 312.23 (a) (6)]

    ☐ b. Investigator data [21 CFR 312.23 (a) (6)(iii)(b)] or completed Form(s) FDA 1572

    ☐ c. Facilities data [21 CFR 312.23 (a) (6)(iii)(b)] or completed Form(s) FDA 1572

    ☐ d. Institutional Review Board data [21 CFR 312.23 (a) (6)(iii)(b)] or completed Form(s) FDA 1572

☐ 7. Chemistry, manufacturing, and control data [21 CFR 312.23 (a) (7)]

    ☐ Environmental assessment or claim for exclusion [21 CFR 312.23 (a) (7)(iv)(e)]

☐ 8. Pharmacology and toxicology data [21 CFR 312.23 (a) (8)]

☐ 9. Previous human experience [21 CFR 312.23 (a) (9)]

☐ 10. Additional information [21 CFR 312.23 (a) (10)]

---

13. IS ANY PART OF THE CLINICAL STUDY TO BE CONDUCTED BY A CONTRACT RESEARCH ORGANIZATION? ☐ YES ☐ NO

    IF YES, WILL ANY SPONSOR OBLIGATIONS BE TRANSFERRED TO THE CONTRACT RESEARCH ORGANIZATION? ☐ YES ☐ NO

    IF YES, ATTACH A STATEMENT CONTAINING THE NAME AND ADDRESS OF THE CONTRACT RESEARCH ORGANIZATION, IDENTIFICATION OF THE CLINICAL STUDY, AND A LISTING OF THE OBLIGATIONS TRANSFERRED.

---

14. NAME AND TITLE OF THE PERSON RESPONSIBLE FOR MONITORING THE CONDUCT AND PROGRESS OF THE CLINICAL INVESTIGATIONS.

---

15. NAME(S) AND TITLE(S) OF THE PERSON(S) RESPONSIBLE FOR REVIEW AND EVALUATION OF INFORMATION RELEVANT TO THE SAFETY OF THE DRUG.

I agree not to begin clinical investigations until 30 days after FDA's receipt of the IND unless I receive earlier notification by FDA that the studies may begin. I also agree not to begin or continue clinical investigations covered by the IND if those studies are placed on clinical hold. I agree that an Institutional Review Board (IRB) that complies with the requirements set forth in 21 CFR Part 56 will be responsible for the initial and continuing review and approval of each of the studies in the proposed clinical investigation. I agree to conduct the investigation in accordance with all other applicable regulatory requirements.

16. NAME OF SPONSOR OR SPONSOR'S AUTHORIZED REPRESENTATIVE	17. SIGNATURE OF SPONSOR OR SPONSOR'S AUTHORIZED REPRESENTATIVE	
18. ADDRESS (Number, Street, City, State and Zip Code)	19. TELEPHONE NUMBER *(Include Area Code)*	20. DATE

(WARNING:A willfully false statement is a criminal offense, USC Title 18, Sec. 1001)

Public reporting burden for this collection of information is estimated to average 30 minutes per response, including the time for reviewing instructions, searching existing data sources, gathering and maintaining the data needed, and completing and reviewing the collection of information. Send comments regarding this burden estimate or any other aspect of this collection of informaiton, including suggestions for reducing this burden to:

Reports Clearance Officer, PHS   and to: Office of Management and Budget
Hubert H. Humphrey Building, Room 721-B   Paperwork Reduction Project (0910-0014)
200 Independence Avenue, S.W.   Washington, DC 20503
Washington, DC 20201
Attn: PRA

                  Please DO NOT RETURN this application to either of these addresses.

Federal regulations state that the "plan should include the following: (a) the rationale for the [product] or the research study; (b) the indication(s) to be studied; (c) the general approach to be followed in evaluating the [product]; (d) the kinds of clinical trials to be conducted in the first year following the submission (if plans are not developed for the entire year, the sponsor should so indicate); (e) the estimated number of patients to be given the [product] in those studies; and (f) any risks of particular severity or seriousness anticipated on the basis of the toxicological data in animals or prior studies in humans with the [biologic] or related [products]" [21 CFR 312.23(a)(3)(iv)].

Investigator's Brochure   The IND must include a complete copy of the investigator's brochure, an information package providing each participating clinical investigator with available information on the biologic, its benefits, and its dangers. The brochure should contain all relevant information that a sponsor has about a biologic and its effects which might prove useful to the investigator in administering the product and monitoring the patient during the trial, including [21 CFR 312.23(a)(5)]:

- a brief description of the biological substance and formulation, including the structural formula, if known;

- a summary of the pharmacological and toxicological effects of the biologic in animals and, to the extent known, in humans;

- a summary of the pharmacokinetics and biological disposition of the biologic in animals and, if known, in humans;

- a summary of information relating to safety and effectiveness in humans obtained from prior clinical studies (reprints of published articles on such studies may be appended when useful); and

- a description of possible risks and side effects anticipated on the basis of prior experience with the biologic under investigation or with related products, and of precautions or special monitoring to be involved in the product's investigational use.

As clinical trials progress, the sponsor must continue to keep investigators informed about the biologic and its effects, particularly adverse effects. Such information may be distributed to investigators by means of periodically revised investigator brochures, reprints or published studies, reports or letters to clinical investigators, or other appropriate means.

Clinical Protocols Clinical protocols are descriptions of clinical studies that identify, among other things, a study's objectives, design, and procedures. Protocols are, of course, a key element of the IND, one which the FDA will evaluate carefully to determine: (1) if Phase 1, 2, or 3 trial subjects will be exposed to any unnecessary risks; and (2) if Phase 2 and 3 study designs are adequate to provide the types and amount of information necessary to show that the drug is safe and/or effective.

In the IND, the sponsor is required to submit only protocols for the study or studies that are scheduled to begin at the end of the FDA's 30-day review period. The safety of the early Phase 1 studies is the FDA's principal concern on receipt of the initial application. Since latter-phase clinical studies are often not fully developed until initial data from Phase 1 studies are obtained, Phase 2 and Phase 3 protocols may be submitted later in the development process.

FDA regulations state that "protocols for Phase 1 studies may be less detailed and more flexible than protocols for Phase 2 and 3 studies. Phase 1 protocols should be directed primarily at providing an outline of the investigation—an estimate of the number of patients to be involved, a description of safety exclusions, and a description of the dosing plan including duration, dose or method to be used in determining dose—and should specify in detail only those elements of the study that are critical to safety, such as necessary monitoring of vital signs and blood chemistries. Modifications of the experimental design of Phase 1 studies that do not affect critical safety assessments are required to be reported to FDA only in the annual report" [21 CFR 312.23(a)(6)(i)].

Protocols for Phase 2 and 3 trials must include detailed descriptions of all aspects of the studies. The regulations state that these protocols "should be designed in such a way that, if the sponsor anticipates that some deviation from the study design may become necessary as the investigation progresses, alternatives or contingencies to provide for such deviations are built into the protocols at the outset. For example, a protocol for a controlled short-term study might include a plan for an early crossover of nonresponders to an alternative therapy" [21 CFR 312.23(a)(6)(ii)].

Biologics Development: A Regulatory Overview

DEPARTMENT OF HEALTH AND HUMAN SERVICES PUBLIC HEALTH SERVICE FOOD AND DRUG ADMINISTRATION **STATEMENT OF INVESTIGATOR** **(TITLE 21, CODE OF FEDERAL REGULATIONS (CFR) PART 312)** *(See instructions on reverse side.)*	Form Approved: OMB No. 0910-0014 Expiration Date: December 31, 1992 See OMB Statement on Reverse. NOTE: No investigator may participate in an investigation until he/she provides the sponsor with a completed, signed Statement of Investigator, Form FDA 1572 (21CFR 312.53(c)).

1. NAME AND ADDRESS OF INVESTIGATOR.

2. EDUCATION, TRAINING, AND EXPERIENCE THAT QUALIFIES THE INVESTIGATOR AS AN EXPERT IN THE CLINICAL INVESTIGATION OF THE DRUG FOR THE USE UNDER INVESTIGATION. ONE OF THE FOLLOWING IS ATTACHED:

        ☐ CURRICULUM VITAE   ☐ OTHER STATEMENT OF QUALIFICATIONS

3. NAME AND ADDRESS OF ANY MEDICAL SCHOOL, HOSPITAL, OR OTHER RESEARCH FACILITY WHERE THE CLINICAL INVESTIGATION(S) WILL BE CONDUCTED.

4. NAME AND ADDRESS OF ANY CLINICAL LABORATORY FACILITIES TO BE USED IN THE STUDY.

5. NAME AND ADDRESS OF THE INSTITUTIONAL REVIEW BOARD (IRB) THAT IS RESPONSIBLE FOR REVIEW AND APPROVAL OF THE STUDY(IES).

6. NAMES OF THE SUBINVESTIGATORS (e.g., research fellows, residents, associates) WHO WILL BE ASSISTING THE INVESTIGATOR IN THE CONDUCT OF THE INVESTIGATION(S).

7. NAME AND CODE NUMBER, IF ANY, OF THE PROTOCOL(S) IN THE IND FOR THE STUDY(IES) TO BE CONDUCTED BY THE INVESTIGATOR.

FORM FDA 1572 (6/92)   PREVIOUS EDITION IS OBSOLETE

8. ATTACH THE FOLLOWING CLINICAL PROTOCOL INFORMATION:

☐ FOR PHASE 1 INVESTIGATIONS, A GENERAL OUTLINE OF THE PLANNED INVESTIGATION INCLUDING THE ESTIMATED DURATION OF THE STUDY AND THE MAXIMUM NUMBER OF SUBJECTS THAT WILL BE INVOLVED.

☐ FOR PHASE 2 OR 3 INVESTIGATIONS, AN OUTLINE OF THE STUDY PROTOCOL INCLUDING AN APPROXIMATION OF THE NUMBER OF SUBJECTS TO BE TREATED WITH THE DRUG AND THE NUMBER TO BE EMPLOYED AS CONTROLS, IF ANY; THE CLINICAL USES TO BE INVESTIGATED; CHARACTERISTICS OF SUBJECTS BY AGE, SEX, AND CONDITION; THE KIND OF CLINICAL OBSERVATIONS AND LABORATORY TESTS TO BE CONDUCTED; THE ESTIMATED DURATION OF THE STUDY; AND COPIES OR A DESCRIPTION OF CASE REPORT FORMS TO BE USED.

9. COMMITMENTS:

I agree to conduct the study(ies) in accordance with the relevant, current protocol(s) and will only make changes in a protocol after notifying the sponsor, except when necessary to protect the safety, rights, or welfare of subjects.

I agree to personally conduct or supervise the described investigation(s).

I agree to inform any patients, or any persons used as controls, that the drugs are being used for investigational purposes and I will ensure that the requirements relating to obtaining informed consent in 21 CFR Part 50 and institutional review board (IRB) review and approval in 21 CFR Part 56 are met.

I agree to report to the sponsor adverse experiences that occur in the investigation(s) in accordance with 21 CFR 312.64.

I have read and understand the information in the investigator's brochure, including the potential risks and side effects of the drug.

I agree to ensure that all associates, colleagues, and employees assisting in the conduct of the study(ies) are informed about their obligations in meeting the above commitments.

I agree to maintain adequate and accurate records in accordance with 21 CFR 312.62 and to make those records available for inspection in accordance with 21 CFR 312.68.

I will ensure that an IRB that complies with the requirements of 21 CFR Part 56 will be responsible for the initial and continuing review and approval of the clinical investigation. I also agree to promptly report to the IRB all changes in the research activity and all unanticipated problems involving risks to human subjects or others. Additionally, I will not make any changes in the research without IRB approval, except where necessary to eliminate apparent immediate hazards to human subjects.

I agree to comply with all other requirements regarding the obligations of clinical investigators and all other pertinent requirements in 21 CFR Part 312.

INSTRUCTIONS FOR COMPLETING FORM FDA 1572
STATEMENT OF INVESTIGATOR

1. Complete all sections. Attach a separate page if additional space is needed.

2. Attach curriculum vitae or other statement of qualifications as described in Section 2.

3. Attach protocol outline as described in Section 8.

4. Sign and date below.

5. FORWARD THE COMPLETED FORM AND ATTACHMENTS TO THE SPONSOR. The sponsor will incorporate this information along with other technical data into an Investigational New Drug Application (IND). INVESTIGATORS SHOULD NOT SEND THIS FORM DIRECTLY TO THE FOOD AND DRUG ADMINISTRATION.

10. SIGNATURE OF INVESTIGATOR	11. DATE

Public reporting burden for this collection of information is estimated to average 1 hour per response, including the time for reviewing instructions, searching existing data sources, gathering and maintaining the data needed, and completing reviewing the collection of information. Send comments regarding this burden estimate or any other aspect of this collection of information, including suggestions for reducing this burden to:

Reports Clearance Officer, PHS
Hubert H. Humphrey Building, Room 721-B
200 Independence Avenue, S.W.
Washington, DC 20201
Attn: PRA

and to:

Office of Management and Budget
Paperwork Reduction Project (0910-0014)
Washington, DC 20503

Please DO NOT RETURN this application to either of these addresses.

FORM FDA 1572 (6/92)

PAGE 2 OF 2

Although the details of a specific protocol will depend upon the phase covered and other factors, the IND regulations specify seven elements that should be included in a protocol [21 CFR 312.23(a)(6)(iii)]:

- a statement of the objectives and purposes of the study;

- the name and address and a statement of qualifications (e.g., resume) of each investigator and the name of each subinvestigator (e.g., research fellow, resident) working under the supervision of the investigator, the names and addresses of the research facilities to be used, and the name and address of each reviewing Institutional Review Board, or IRB (this information may be submitted on Form FDA 1572—see sample above);

- the criteria for patient selection and for exclusion of patients and an estimate of the number of patients to be studied;

- a description of the design of the study, including the kind of control group to be used, if any, and a description of the methods to be used to minimize bias on the part of subjects, investigators, and analysts;

- the method for determining the dose(s) to be administered, the planned maximum dosage, and the duration of individual patient exposure to the biologic;

- a description of the observations and measurements to be made to fulfill the objectives of the study; and

- a description of clinical procedures, laboratory tests, or other measures to be taken to monitor the effects of the biologic in human subjects and to minimize risk.

Chemistry, Manufacturing, and Control Information The sponsor must establish in the IND that it can manufacture the biologic while ensuring adequate levels of product identity, quality, purity, strength, and stability. The FDA requires that the IND's chemistry, manufacturing, and control information section consist of five elements [21 CFR 312.23(a)(7)(ii)]:

Biological Substance. "A description of the [biological] substance, including at least five general elements: its physical, chemical or biological characteristics; the name and address of its manufacturer; the general method of preparation of the substance; the acceptable limits and analytical methods used to assure the identity, strength, quality and purity of the substance; and information sufficient to support the stability of the substance during toxicological studies and the planned clinical studies. Reference to the current edition of the U.S. Pharmacopeia or National Formulary may satisfy relevant requirements in this paragraph."

Biological Product. "A list of all components, which may include reasonable alternatives for inactive compounds, used in the manufacture of the investigational ... product and those which may not appear but which are used in the manufacturing process and, where applicable, the quantitative composition of the investigational ... product, including any reasonable variations which may be expected during the investigational stage; the name and address of the ... product manufacturer; a brief general description of the manufacturing and packaging procedure for the product; the acceptable limits and analytical methods used to assure the identity, quality, purity and strength of the ... product; and information sufficient to assure the product's stability during the planned clinical studies. Reference to the current edition of the U.S. Pharmacopeia or National Formulary may satisfy certain requirements in this paragraph."

A Brief General Description of the Composition, Manufacture, and Control of any Placebo Used in a Controlled Clinical Trial. This requirement is designed to ensure that the failure of a placebo to mimic the odor, taste, texture, or other physical characteristics of an investigational product does not compromise a blinded study. The FDA does not require that a placebo be identical in all aspects to an investigational product.

Labeling. A copy of all labels and labeling to be provided to each investigator.

Environmental Analysis Requirements. Either an environmental assessment or a claim for a categorical exclusion from the requirement for an environmental assessment is required. In INDs, categorical exclusions are more common.

There are specific regulations for claiming exclusion and for the contents of an environmental assessment (see Chapter 9).

The FDA has published a number of guidelines to assist sponsors in submitting chemistry, manufacturing, and controls information. In addition, CBER has several guidelines and points-to-consider documents for the identification, testing, and manufacturing of specific types of biological products. These documents are available through CBER (for a complete listing of these and other CBER guidance documents, see Appendix 1).

The detail of information needed in the chemistry, manufacturing, and controls section depends on several factors, including the scope of the clinical investigation, the proposed duration of clinical testing, and the product's dosage form. The IND regulations state that "the emphasis in an initial Phase 1 submission should generally be placed on the identification and control of the raw materials and the new [biological] substance. Final specifications for the biological substance and [biological] product are not expected until the end of the investigational process" [21 CFR 312.23(a)(7)(i)]. However, when final specifications are not established until just prior to Phase 3 studies, comparability with preceding studies may be necessary. Stability testing should support the duration of use for the proposed clinical study(s).

The FDA does require that sponsors comply with Current Good Manufacturing Practices (cGMP) during clinical trials. According to the FDA's *Guideline on the Preparation of Investigational New Drug Products (Human and Animal)* (March 1991): "FDA recognizes that manufacturing procedures and specifications will change as clinical trials advance. However, as research nears completion, procedures and controls are expected to be more specific because they will have been based upon a growing body of scientific data and documentation.... When [product] development reaches a stage where the [biologics] are produced for clinical trials in humans ... then compliance with the cGMP regulations is required. For example, the [biological] product must be produced in a qualified facility, using laboratory and other equipment that has been qualified, and processes must be validated. There must be written procedures for sanitation, calibration and maintenance of equipment, and specific instructions for the use of the equipment and procedures used to manufacture the [biologic]. Product contamination and wide variations in potency can produce substantial levels of side effects and toxicity, and even produce wide-sweeping effects on the physiological activity of the [product]. Product safety, quality, and uniformity are especially significant in the case of investigational products. Such factors may affect the outcome of a clinical investigation that will, in large part, determine whether or not the product will be approved for wider distribution to the public."

Pharmacology and Toxicology Information   The sponsor must provide information from pharmacological and toxicological animal studies sufficient to show that the clinical studies proposed in the IND will not expose human subjects to unreasonable risks. The preclinical testing requirements applicable to the biologic will depend on the type of product, its use and length of administration, and other factors. In this section of the IND, the following three elements should be included [21 CFR 312.23(a)(8)]:

- pharmacology and biologic disposition, which describes the [product's] pharmacological effects and mechanism of action, and, if known, information on the absorption, distribution, metabolism, and excretion of the biologic;

- toxicology, which provides an integrated summary of the toxicological effects of the biologic in animals and *in vitro*, and a full tabulation of data for each toxicology study that is intended primarily to support the safety of the proposed clinical trial;

- either a statement that each nonclinical laboratory study subject to Good Laboratory Practice (GLP) regulations was conducted in adherence to these standards, or, if not, a statement explaining why GLP standards were not followed.

Adequate information to support clinical studies will vary from minimal to substantial, depending on the nature of the product and the proposed indication (see Chapter 2). Discussions with CBER on product-specific requirements are encouraged early in product development.

Previous Human Experience with the Investigational Product
If an investigational product, or any of its active ingredients, has been marketed

or tested in humans previously, the sponsor must provide specific information about any data that may be relevant to the FDA's evaluation of product safety.

Additional Information   Regarding INDs for products that are radioactive or that contain ingredients with a potential for abuse, FDA regulations encourage the submission of other information that might be relevant to the proposed clinical studies. Additional information that would assist the agency in evaluating the IND with regard to safety, trial design, or the use of a study to support a marketing application may be included in this section.

Relevant Information   The FDA has the option of requiring any other information that it believes is relevant to the safety review of any biologic and clinical trial program.

The IND and FDA-Sponsor Communications

To provide advice or request information, the FDA may contact the sponsor by letter or telephone. The agency may provide advice to sponsors on its own initiative or in response to a sponsor request. Unless an IND is on clinical hold (see Chapter 5), modifications suggested by FDA reviewers are not considered formal requirements. Sponsors are strongly encouraged to follow such recommendations, however.

At various points in a product's development, meetings between the FDA and the sponsor may be appropriate to address questions and issues. One stage at which such a meeting might be appropriate for important and novel products is just prior to the IND submission. Such meetings, sometimes called "pre-IND meetings," allow sponsors to obtain FDA recommendations on the IND submission and give agency reviewers a preview of the upcoming application and associated issues. Sponsors must submit background information prior to such meetings.

Maintaining the IND

The IND is a "living document." After the initial IND submission, the FDA requires that IND sponsors continually update their applications with information that allows the agency to reassess periodically the safety of ongoing and future clinical trials. Federal regulations require sponsors of active INDs —those not having been formally discontinued in any way (see Chapter 5)— to file four types of updates to their applications: protocol amendments, information amendments, IND safety reports, and annual reports.

These submissions and all other correspondence submitted regarding an IND are considered amendments. To facilitate the document handling and review process, CBER encourages sponsors to attach to all amendments Form FDA 1571, and to sequentially number section, or block, 10 (serial number) of that form.

Although IND amendments should be submitted as necessary, sponsors are discouraged from submitting amendments more frequently than every 30 days.

An amendment may contain many types of information. For example, an amendment might include a new protocol along with additional pharmacology information. The amendment should be clearly identified with the amendment types. This is most easily achieved by submitting Form FDA 1571 and checking the appropriate box(es) in block 11. The submission should also be organized in a manner that helps reviewers locate the various sections of information in the amendment.

Protocol Amendments   Clinical protocol amendments are submitted when a sponsor wants to make a change to a previously submitted protocol, add an investigator to a previously submitted protocol, or add a study protocol not submitted in the original IND. New protocols and most protocol changes must be submitted to the FDA and receive IRB approval before being initiated.

Amendments providing for a new protocol must contain the protocol itself along with "a brief description of the most clinically significant differences" between the new and previously submitted protocols [21 CFR 312.30(d)(i)]. This analysis facilitates CBER's review of major changes (e.g., changes in dosing, route of administration, indication) that may require additional supporting data. Sponsors may implement new protocols upon submission to the FDA, provided that IRB approval has been obtained. Some sponsors may choose to obtain FDA comments before implementing new protocols for pivotal studies.

Amendments filed for changes to previously submitted protocols must include a description of any change in a Phase 1 protocol that may significantly affect the safety of subjects, or any change in a Phase 2 or 3 protocol that may significantly affect the safety of the subjects, the scope of the investigation, or the scientific quality of the study. The IND regulations provide several examples of protocol changes requiring amendments [21 CFR 312.30(b) and (c)]:

- any increase in biologic dosage or duration of exposure of individual subjects to the product beyond that in the current protocol, or any significant increase in the number of subjects under study;

- any significant change in the design of a protocol, such as the addition or deletion of a control group;

- the addition of a new test or procedure that is intended to improve monitoring for, or reduce the risk of, a side effect or adverse effect, or the elimination of a test intended to monitor safety;

- a protocol change intended to eliminate an apparent hazard to subjects (these changes may be implemented prior to an amendment submission, provided that the FDA is subsequently notified through the protocol amendment and that the IRB is properly notified); and

- the addition of a new investigator to carry out a previously submitted protocol (these changes may be implemented before the amendment submission—the investigational product may be shipped to the investigator and the investigator may participate in the study, provided the sponsor notifies the FDA within 30 days of the investigator's first participation in the study).

For an amendment changing a previously submitted protocol, a cover letter noting the specific changes or an annotated copy of the protocol indicating the changes will facilitate the review.

One particularly important requirement regarding protocol amendments calls for sponsors to reference the specific information relied upon to support the new protocol or protocol change. According to the regulations, a protocol amendment must contain a "... reference, if necessary, to specific technical information in the IND or in a concurrently submitted information amendment to the IND that the sponsor relies on to support any clinically significant change in the new or amended protocol. If the reference is made to supporting information already in the IND, the sponsor shall identify by name, reference number, volume, and page number the location of the information" [21 CFR 312.30(c)(2)]. Thus, if a sponsor intends to change the dosage form of the investigational product, appropriate animal tests that would support this increased human exposure are required. To the extent that the FDA is apprised of the basis for a change in a protocol, it can move quickly and comprehensively to review the change. Of course, if the change is one that plainly does not require specific technical support, the sponsor would not be expected to provide a reference.

Information Amendments   IND information amendments are used to report to the FDA important new information that would not ordinarily be reported in a protocol amendment, IND safety report, or annual report. New or additional pharmacology, toxicology, chemistry, manufacturing, or control data or other technical information, and reports of discontinuation of clinical trials are commonly reported in information amendments.

Content requirements for information amendments are: "(1) a statement of the nature and purpose of the amendment, (2) an original submission of the data in a format appropriate for scientific review, (3) if the sponsor desires FDA to comment on an information amendment, a request for such comment" [21 CFR 312.31 (a) and (b)].

IND Safety Reports   The FDA has extremely specific requirements for the processing and reporting of clinical and nonclinical adverse reactions. The agency's goal in this area is to ensure timely communication about experiences with the investigational product.

The IND regulations require sponsors to review all information that might represent a possible adverse reaction: "Sponsors must promptly review all information relevant to the safety of the [product] from any source, foreign or domestic, including information derived from clinical investigations, animal investigations, commercial marketing experience, reports in the scientific literature, and unpublished scientific papers" [21 CFR 312.32(b)]. Once an employee of a company has knowledge of safety data, the sponsor is considered to have received safety information.

Reporting requirements for any adverse experience will depend directly on the nature, severity, probable cause, and frequency of the experience. To determine reporting requirements, the following definitions are important:

Serious Adverse Experience. A serious adverse event is "any experience that suggests a significant hazard, contraindication, side effect, or precaution." Under its new MedWatch Program, the FDA has proposed a revised definition of "serious" clinical adverse events, and encouraged industry to begin employing the new definition immediately: "an adverse experience occurring at any dose that is fatal or life-threatening, results in persistent or significant disability/incapacity, requires or prolongs inpatient hospitalization, necessitates medical or surgical intervention to preclude impairment of a body function or permanent damage to a body structure, or is a congenital anomaly. With regard to results obtained from tests in laboratory animals, a serious adverse ... experience includes any experience suggesting a significant risk for human subjects, including any finding of mutagenicity, teratogenicity, or carcinogenicity" [21 CFR 312.32(a)].

Unexpected Adverse Experience. An unexpected adverse event is "any adverse experience that is not identified in nature, severity, or frequency in the current investigator's brochure" [21 CFR 312.32(a)]. If an investigator's brochure is not required, this refers to any adverse experience that is not identified in nature, severity, or frequency in the risk information specified in the general investigational plan or elsewhere in the IND or its amendments.

For an unexpected fatal or life-threatening experience associated with the use of a biologic in clinical studies conducted under an IND, the sponsor is required to notify the FDA initially through a telephone report made within three working days after the sponsor receives information on the reaction. The phrase "associated with the use of the drug" is interpreted by the regulations to mean that there is a reasonable possibility that the experience may have been caused by the product [21 CFR 312.32(a) and (c)(2)].

Written IND safety reports are required for any clinical or preclinical "adverse experience associated with the use of the [product] that is both serious and unexpected." These reports must be submitted to the FDA no later than ten working days after the sponsor's initial receipt of the information.

Also, the reports must identify "all safety reports previously filed with the IND concerning a similar adverse experience and analyze the significance of the adverse experience in light of the previous, similar reports." Adverse reactions that require the three-day telephone alert must also be the subject of a written safety report [21 CFR 312.32(c)(1)].

In cases in which ten working days are not sufficient to determine conclusively whether an adverse event should be reported, the agency advises that sponsors submit preliminary information and supplement this initial report with whatever more definitive information is subsequently obtained. An exception might be the most serious adverse experiences obtained from non-clinical studies. A report should be filed after a complete analysis reveals that the experimental product caused that reaction.

Successive cases of serious and unexpected adverse reactions must be reported in written reports until the risk posed by the experience is sufficiently well understood to be described in the investigator brochure or until an equally satisfactory resolution of the issue is reached (for example, a determination that the experience is not product related).

Sponsors should remember that a safety report filed to an IND does not necessarily represent a concession that there is a relationship between the product and the adverse experience.

Annual Reports   Within 60 days of the anniversary date on which the initial IND was either allowed to proceed or placed on clinical hold, the sponsor must submit an overview of information collected during the previous year for the subject product(s). The report should cover the following areas [21 CFR 312.33]:

Information on Individual Studies. The FDA wants "a brief summary of the status of each study in progress and each study completed during the previous year. The summary must include the following information on each study: (1) the title of the study (with any appropriate study identifiers such as protocol number), its purposes, a brief statement identifying the patient population and a statement as to whether the study is completed; (2) the total number of subjects initially planned for inclusion in the study, the number enrolled in the study to date, the number whose participation in the study was completed as planned, and the number who dropped out of the study for any reason; and

(3) if the study has been completed or if the interim results are known, a brief description of any available study results."

Summary Information. This section should include all additional information collected pertaining to the product, as well as summary data from all the clinical studies:

- a narrative or tabular summary showing the most frequent and most serious adverse experiences by body system;

- a summary of all the IND safety reports submitted during the past year;

- a list of subjects who died during participation in the investigation, with the cause of death for each subject (this list must include all deaths, including those persons whose cause of death is not believed to be product related);

- a brief description of any information that is pertinent to an understanding of the biologic's actions, including, for example, information from controlled trials, and information about bioavailability;

- a list of the preclinical studies (including animal studies) completed or in progress during the past year and a summary of the major preclinical findings; and

- a summary of any significant manufacturing or microbiological changes made during the past year.

General Investigational Plan. A brief description of the general investigational plan for the coming year must be submitted. This plan must be as descriptive as that submitted in the original IND. If the plans are not yet formulated, the sponsor must indicate this fact in the report.

Investigator's Brochure Revisions. When the investigator's brochure has been revised, the sponsor must include a description of the revision and a copy of the new brochure.

Phase 1 Modifications. The sponsor must describe any significant Phase 1 protocol modifications made during the previous year and not previously reported to the FDA through a protocol amendment.

Foreign Marketing Developments. According to the IND regulations, this section should provide a "brief summary of significant foreign marketing developments with the [biologic] during the past year, such as approval of marketing in any country or withdrawal or suspension from marketing in any country."

Request for an FDA Response. If the sponsor requests an FDA meeting, reply, or comment, a log of any relevant outstanding business with respect to the IND must be included.

This chapter reflects the author's assessment of the requirements for IND submissions and is not intended to represent the official position of the FDA.

Chapter 5
CBER and the Biological IND Review

by Mark Mathieu
PAREXEL International Corporation

When a sponsor submits an IND, the FDA assumes an important role in the development of a new biological product. From this point forward, the sponsor can do little without at least submitting documents to, and possibly waiting for a review and approval from, the FDA. This fact is emphasized in the statement that a sponsor must sign in the IND (Form FDA 1571): "I agree not to begin clinical investigations until 30 days after FDA's receipt of the IND unless I receive earlier notification by FDA that the studies may begin."

Companies developing biological products interact with the FDA primarily through CBER. Therefore, before outlining the biological IND review process, it is appropriate to first profile CBER, the regulatory and scientific unit responsible for processing and evaluating biological INDs, for reviewing and approving PLAs and ELAs, and for all other product review and compliance issues regarding biologics.

(Editor's note: Although the term "biological IND" is used throughout this chapter, CBER sometimes reviews INDs for drugs and investigational device exemptions (IDE) for medical devices. In most cases, CBER's review processes for, and approach to, these applications are similar—if not identical—to those for biological INDs.)

The FDA's Center for Biologics Evaluation and Research (CBER)

In early 1993, CBER virtually reinvented itself. As part of a much-publicized reorganization, CBER created a new management and review structure designed to streamline product reviews, and to allow the center to better accommodate a growing number of applications for biotechnology products.

During that period, however, the restructuring was only one source of change for CBER. The passage of the Prescription Drug User Fee Act of 1992 promised to provide funding for an additional 300 staffers and established, for the first time, a statutory "deadline" for product and establishment license application reviews. Meanwhile, CBER's senior management had developed several initiatives, including a new "managed" review process and tighter standards for license applications, designed to improve the biologic approval process, upgrade CBER's information and project management capabilities, and eliminate unnecessary regulatory burdens on industry. Collectively, the reorganization, new legislation, and internal CBER initiatives were expected to bring a new era of biological regulation.

While this chapter focuses primarily on CBER's IND review processes, it is worth noting that the reorganization and related initiatives brought profound change to the review structure and processes for PLAs and ELAs as well. Perhaps most importantly, the changes centralized IND and PLA reviews into therapeutic area-based review and research divisions, and made biologic approval processes more consistent with those for drugs (for more on PLA and ELA reviews, see Chapter 10).

CBER's Recent History   CBER was created in late 1987 as part of an FDA reorganization plan to separate the reviews of drugs and biologics. The reorganization split the FDA's Center for Drugs and Biologics into two distinct centers: CBER and CDER. FDA officials believed that this new structure provided more definitive lines of authority for drug and biologic reviews.

In the early 1990s, however, CBER officials recognized that new technologies such as gene and somatic cell therapy had, or were about to, present the center with new scientific and regulatory challenges. Of principal concern was CBER's workload, which increased substantially due primarily to a wave of biotechnology products entering the clinical testing stage. From 1989 to 1992, for example, IND submissions for biotechnology products more than doubled, to more than 300 in 1992 (see chart below). And in both 1991 and 1992, biotechnology INDs represented well over half the INDs submitted to CBER.

Perhaps more troubling for CBER were the implications this nascent trend seemed to hold for the mid-1990s. Ultimately, the biotechnology research reflected by IND submissions will produce a new generation of PLA submissions, which require more resource-intensive reviews than INDs. Such prospects concerned FDA officials, who were keenly aware that CBER's product review pipeline was already under strain: the important new biological products approved in 1992 had an average review time of 35.3 months, according to the U.S. Pharmaceutical Manufacturers Association (PMA).

In 1992, CBER officials developed a restructuring plan to help the center address these new demands. In doing so, the officials defined at least six major objectives for the reorganization: (1) to optimize and streamline the biologics review process; (2) to delegate authority and responsibility lower in CBER's organization; (3) to maintain and build strong scientific programs; (4) to better integrate research and review functions; (5) to maintain and build a strong and vigilant postmarketing compliance program; and (6) to ensure clear, consistent, and effective communication within the agency and with other agencies, industry, the public, and the U.S. Congress.

None of these initiatives changed CBER's overall responsibilities or mission. CBER continues to define its mission as the "Protection and enhancement of the public health of this nation in the areas of blood, vaccines and biological therapeutics."

CBER's New Structure

The 1993 reorganization brought fundamental change to CBER's structure for product reviews. Prior to the reorganization, CBER employed a structure segregated largely by activities, such as IND reviews, PLA reviews, laboratory research, and ELA reviews. A product's IND review, for instance, would be led or coordinated by a different division than the PLA for the same product. Some believe that this structure led to reviews that were, in some ways, inefficient and fragmented, and that unnecessarily isolated product reviews from CBER's highly regarded laboratory research staffers, who could bring to the review process insights provided by their cutting-edge research activities.

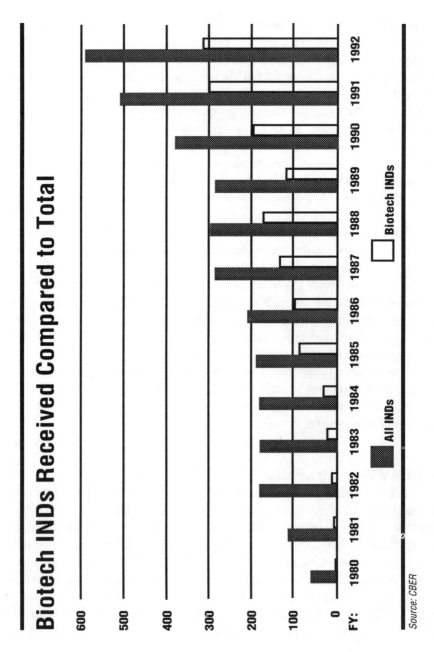

Chapter 5: CBER and the Biological IND Review

The structure that CBER assumed officially on January 25, 1993, is designed to eliminate review inefficiencies, and to concentrate, in three product-specific offices, most research and product review functions (see chart below). These three offices—the Office of Vaccines Research and Review, the Office of Therapeutics Research and Review, and the Office of Blood Research and Review—will review all biologic INDs and PLAs, and conduct all laboratory research in their respective areas of expertise and responsibility.

The concentration of the administrative, medical, regulatory, and scientific expertise necessary for IND and PLA reviews is the most fundamental change brought by CBER's reorganization. This is expected to add greater continuity to the biologic development and review process, since sponsors will deal with a single product review group throughout the product development cycle. However, some aspects of the product review process—ELA and statistical reviews, for example—will be lead by groups located elsewhere within CBER.

By combining several groups that were previously loosely affiliated, CBER's reorganization is forcing CBER officials and staffers to relate and function differently in some respects. The adjustment process is expected to consume both time and resources. For example, CBER staffers that had been involved only with INDs had to begin "cross-training" on PLA issues as well.

Under the reorganization, CBER's product-specific offices are flanked by three offices that provide expertise and support services in several key areas. The center's six offices are profiled below.

The Office of Vaccines Research and Review The Office of Vaccines Research and Review evaluates all INDs and PLAs for, and conducts laboratory research on, vaccine, anti-bacterial, and allergenic products. Staffed largely with scientists, three product-specific divisions conduct all product research, and provide scientific and technical support for IND and PLA reviews: the Division of Allergic Products and Parasitology; the Division of Bacterial Products; and the Division of Viral Products.

According to a 1993 internal CBER document, *Guide List for Product Assignments to Offices*, these three divisions will review and conduct research on the following products, among others: allergenic extracts, bacterial products, BCG (alone), L-asparaginase, skin tests, snake venoms, toxins, toxoids, and vaccines (not tumor).

CENTER FOR BIOLOGICS EVALUATION AND RESEARCH

Director

OFFICE OF MANAGEMENT

OFFICE OF COMPLIANCE
- Division of Case Management
- Division of Bioresearch Monitoring and Regulations
- Division of Inspection and Surveillance

OFFICE OF BLOOD RESEARCH AND REVIEW
- Division of Transfusion Transmitted Diseases
- Division of Hematology
- Division of Blood Collection and Processing
- Division of Blood Establishment and Product Applications

OFFICE OF THERAPEUTICS RESEARCH AND REVIEW
- Division of Cytokine Biology
- Division of Cellular & Gene Therapies
- Division of Hematologic Products
- Division of Monoclonal Antibodies
- Division of Clinical Trial Design and Analysis
- Division of Application Review and Policy

OFFICE OF VACCINES RESEARCH AND REVIEW
- Division of Allergic Products and Parasitology
- Division of Bacterial Products
- Division of Viral Products
- Division of Vaccine and Related Products Applications

OFFICE OF ESTABLISHMENT LICENSING AND PRODUCT SURVEILLANCE
- Division of Veterinary Services
- Division of Establishment Licensing
- Division of Biostatistics and Epidemiology
- Division of Product Quality Control

The Office of Vaccines Research and Review has a fourth division that fulfills a key role in IND and PLA reviews—the Division of Vaccine and Related Products Applications (DVRPA). Staffed by scientists, physicians, and consumer safety officers (CSO), the division leads and coordinates application reviews, and serves as sponsors' point of contact with the office.

The Office of Therapeutics Research and Review In many ways, CBER's new Office of Therapeutics Research and Review is one of the more intriguing offices within the restructured center. Perhaps because it stands to bear much of CBER's growing workload—more than 60 percent of 1992 IND submissions were for biological therapeutics (see chart below)—the office has some unique structural features.

The Office of Therapeutics Research and Review has six functioning divisions, including four product-specific divisions that conduct laboratory research and provide scientific and technical support for IND and PLA reviews: the Division of Cytokine Biology; the Division of Cellular and Gene Therapies; the Division of Hematologic Products; and the Division of Monoclonal Antibodies.

According to CBER's product assignment list, these divisions will review and conduct research on the following products, among others:

- alteplase
- BCG with tumor cells
- bone marrow
- CD4
- ciliary neuro. factor
- colony stimulating factor
- cultured lymphocytes
- erythropoietin
- G-CSF
- GM-CSF
- interferon
- keyhole lim. hemocyanin (KLH)
- leukocyte dialysate
- mobilized blood cells
- platelet lysate

- ancrod
- beta glucan polymer
- carboxypeptidase
- cell sep./purg. devices
- collagenase
- corynebacterium parvum
- DNase
- fibrinogen
- gene therapy
- growth factors
- interleukin
- M-CSF
- monoclonal antibodies
- protein C
- somatic cell therapy

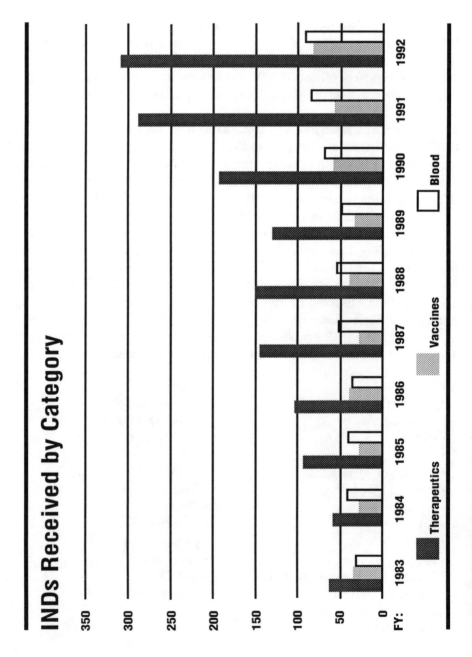

- retroviral vectors
- stem cell factor
- tissue plasminogen activator (t-PA)
- tumor necrosis factor
- urokinase

- streptokinase
- thrombolysin
- transfer factors
- tumor vaccines

Like DVRPA, the Division of Application Review and Policy (DARP) handles the administrative and regulatory processing of submissions, and leads or coordinates IND and PLA reviews. Staffed largely with scientists and CSOs, DARP coordinates sponsor communications and ensures consistency in regulatory decision-making within the Office of Therapeutics Research and Review.

The Division of Clinical Trial Design and Analysis is a group unique to the Office of Therapeutics Research and Review. As implied by its name, the division provides the office's research and review divisions with expertise in clinical trial methodologies and analyses. In doing so, the division works closely with statisticians from the Division of Biostatistics and Epidemiology. The unit is staffed largely by physicians, who consult on issues such as clinical protocols, clinical data reviews, and the evaluation of post-approval adverse experiences. The division also houses a pharmacology/toxicology group, which provides expertise in nonclinical testing and data analysis, and CBER's Oncology Products Branch, which works with CDER's Division of Oncology and Pulmonary Products on reviews of certain cancer therapies.

The Office of Blood Research and Review   The Office of Blood Research and Review is unique among CBER's three product research and review offices in that it reviews ELAs in addition to INDs and PLAs. Because of the special expertise needed for inspections of plasma donor centers and blood banks, the office leads inspections of these establishments.

In most other respects, the Office of Blood Research and Review is much like its two counterparts. The unit consists of four divisions, including three product-specific divisions that conduct laboratory research and provide technical expertise for IND, PLA, and certain ELA reviews: the Division of Transfusion Transmitted Diseases; the Division of Hematology; and the Division of Blood Collection and Processing.

According to CBER's product assignment list, these three divisions review and conduct research on the following products, among others:

- albumin
- antithrombin
- blood banking products
- blood products
- devices (not cell sep.)
- dextran
- Factor 4
- Factor VII
- fibrin sealant
- fibronectin
- hematin
- hetastarch
- immune globulin
- immune plasma
- *in vitro* test kits
- pentastarch
- polyclonal antibodies
- sec. leuk. prot. inhib.
- succinylated gelatin
- von Willebrand frag.
- antihemophilic factor
- apolipoprotein-A
- blood group substances
- colostrum
- dextrose solutions
- Factor IX
- Factor VIII
- fibrinogen frag. E
- fluosol
- heme arginate
- hydroxyethyl starch
- immune milk
- immune whey
- pentaglobin
- perfluorodecali
- proteinase inhibitor
- snake anti-venoms
- thrombi
- WinRho

The Division of Blood Establishment and Product Applications (BEPA) leads or coordinates IND, PLA, and ELA reviews, and handles all administrative and regulatory processing of these applications. The division also manages all sponsor communications regarding submissions.

The Office of Establishment Licensing and Product Surveillance
As the office responsible for most ELA reviews, CBER's Office of Establishment Licensing and Product Surveillance is also fundamentally important to the product review process. In providing biostatistical expertise to the IND and PLA review units and managing a postmarketing surveillance program as well, the office's four divisions play a variety of roles during and following the product review process:

Division of Establishment Licensing. The division reviews all ELAs, except those for plasma donor centers and blood banks.

Division of Product Quality Control. The division manages CBER's biologics samples testing program for lot-by-lot release authorization.

Division of Biostatistics and Epidemiology. This unit provides statistical expertise to CBER's research and product review divisions. The unit also manages CBER's postmarketing product surveillance program.

Division of Veterinary Services. The division provides animal care, use, and pathology services and technical assistance. It also manages CBER's Animal Care Committee, which reviews all animal testing protocols for CBER's research activities.

The Office of Establishment Licensing and Product Surveillance also houses CBER's advertising, promotion, and labeling staff. Although not established as a separate division, the group reviews all advertising and promotional labeling for biological products (see Chapter 12).

The Office of Compliance The Office of Compliance may be the CBER unit least affected by the center's reorganization. Much as it did prior to the restructuring, the office fulfills its monitoring and compliance responsibilities through three divisions:

Division of Bioresearch Monitoring and Regulations. The division manages CBER's Bioresearch Monitoring Program, which inspects clinical and nonclinical research programs to ensure compliance with FDA standards.

Division of Inspection and Surveillance. The unit manages CBER's prelicensure and annual manufacturing inspection program for all licensed establishments. The division also tracks and reviews all manufacturing compliance actions and manufacturing errors and accident reports.

Division of Case Management. The division coordinates all CBER compliance actions, including the implementation of product recalls, the issuance of regulatory letters, and the revocation of product and establishment licenses.

The Office of Management CBER's reorganization did not affect the structure or roles of the center's Office of Management, which spearheads the center's planning, budgeting, training, and information management activities.

CBER's Review Process for INDs

Although CBER's restructuring and the related legislative and management initiatives affected most center activities, they may have brought less fundamental changes to the IND review process than to the PLA/ELA review process. Still, the reorganization has affected IND reviews, which are expected to evolve further as CBER refines its procedures under the new structure.

The principal regulatory goals of, and scientific criteria applied during, IND reviews remain unaffected, however. During IND reviews, CBER attempts: (1) to determine if the available research data show that the product is reasonably safe for administration to human subjects; and (2) to determine if the protocol for the proposed Phase 1 clinical studies will expose subjects to unnecessary risks.

CBER's restructuring has reshaped the review path for INDs. Previously, INDs for all products were funnelled to the Division of Biological INDs, which led and coordinated all IND reviews. Under CBER's new structure, IND review responsibilities have been segregated based on product type.

All biological INDs begin the review process in CBER's Document Control Center, where staffers log in certain information about the submission—sponsor name, receipt date, and product name. Most importantly at this point, the IND receives CBER's receipt stamp, which starts the 30-day period within which CBER must reach a decision on the IND.

After CBER's Document Control Center completes its initial processing, the paths of INDs diverge. Based on the type of product for which the application is filed, the IND is distributed to one of three product area-based research and review offices discussed above: the Office of Vaccines Research and Review; the Office of Therapeutics Research and Review; or the Office of Blood Research and Review. While the structures and review procedures of the three offices differ in some ways, the IND review process is, in most important respects, similar across these units.

A Profile of CBER's IND Review Processes

After determining which of CBER's three product research and review offices has authority over an IND, the center's Document Control Center forwards the submission to that office's product application division:

- DVRPA, within the Office of Vaccines Research and Review (see diagram below);

- DARP, within the Office of Therapeutics Research and Review (see diagram below); or

- BEPA, within the Office of Blood Research and Review (see diagram below).

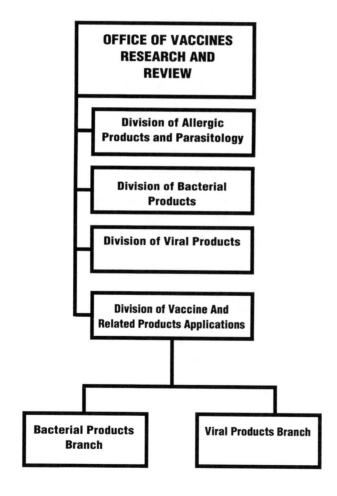

Biologics Development: A Regulatory Overview

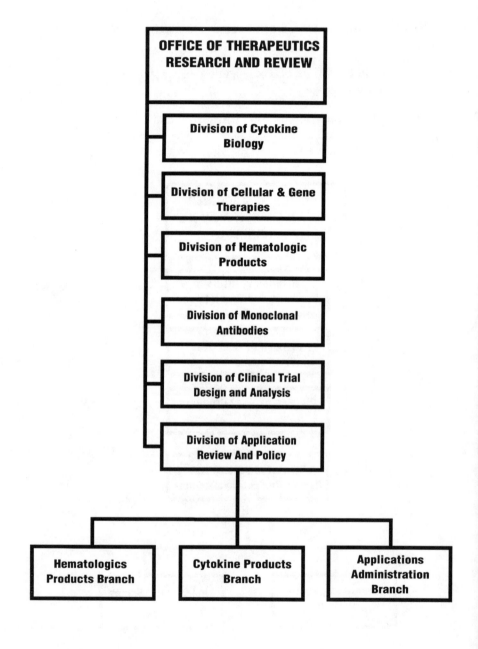

90

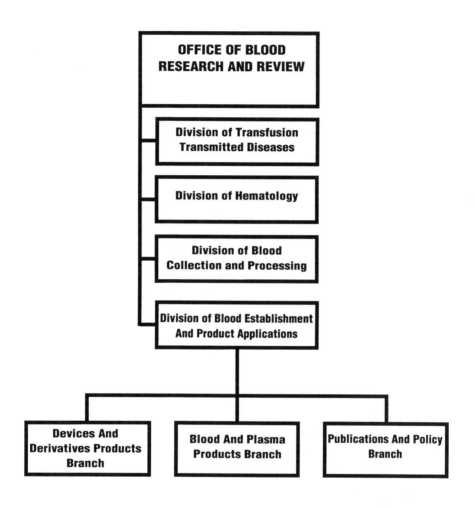

Once within the relevant division, the IND is assigned to a CSO, who first conducts an administrative review of the application to ensure: (1) that the IND contains sufficient information to justify a scientific review; (2) that the submission contains all necessary completed forms; (3) that the product belongs under the authority of CBER and the division to which the application has been forwarded; and (4) that all regulatory references are correct.

If the IND is deficient in some respect (e.g., the submission is missing a form or the investigator's brochure), the CSO may telephone the sponsor about the deficiency. If the sponsor can remedy the problem quickly, the IND review may proceed and a clinical hold order can be averted (for a complete discussion of clinical holds, see below). For less serious IND deficiencies, the CSO may decide to forward the application for review and notify the sponsor of the need to address the deficiencies.

Provided the IND fulfills the basic submission requirements, the CSO prepares an acknowledgement letter that informs the sponsor of the IND's application number and the sponsor's agency contact person within BEPA, DVRPA, or DARP. Perhaps most importantly, the acknowledgement letter identifies the date on which CBER received the IND, information that will tell the sponsor when it can initiate clinical trials, provided the FDA does not intervene.

After evaluating what types of expertise are necessary to review the IND, either the CSO or a branch chief within the division then selects the reviewers to evaluate the application. In most cases, the IND is assigned to a "primary reviewer," a scientist or physician within one of the branches within BEPA, DVRPA, or DARP:

- DVRPA has two branches—the Bacterial Products Branch and the Viral Products Branch.

- BEPA has three branches—the Devices and Derivatives Products Branch, the Blood and Plasma Branch, and the Publications and Policy Branch (as late as mid-1993, BEPA had neither staffed this unit nor released details about its responsibilities).

- DARP has three branches—the Hematologics Products Branch, the Cytokine Products Branch, and the Applications Administration Branch, a unit that houses all CSOs within DARP and that handles all initial application processing.

After assessing the expertise necessary for the review and logging in certain information about the IND and its sponsor, the CSO then returns the IND to CBER's Central Document Room, which is responsible for distributing the application to all assigned reviewers. Before forwarding the application, however, Central Document Room staffers bind the IND copies and log in pertinent IND-related information for application-tracking purposes.

Typically, the BEPA, DVRPA, or DARP reviewer assigned to the application acts as the "primary" IND reviewer, meaning that he or she will lead the IND evaluation process. In many cases, the primary reviewer obtains consultative reviews from scientists and/or physicians within one or more of the office's research and review divisions. Within the Office of Vaccines Research and Review, for example, a scientist or physician from either the Division of Allergic Products and Parasitology, the Division of Bacterial Products, or the Division of Viral Products might be assigned to work on the IND. In some situations, one of these researchers may serve as the primary IND reviewer.

Much of CBER's reputation for having highly qualified and informed reviews is related to the strong science base provided by these laboratory research and review divisions. Because they are staffed with physicians and scientists who are involved in both regulatory decision-making and laboratory research, these units are credited with possessing unique expertise to review even the most scientifically advanced products.

The number of reviewers assigned to an IND depends on several factors, including the complexity of the specific product under consideration. The review assignments also differ within CBER's product review offices. Within the Office of Therapeutics Research and Review, for example, an IND's proposed clinical protocols are reviewed by physicians within the Division of Clinical Trial Design and Analysis. In the other two offices, clinical protocols are typically evaluated by the primary reviewer assigned to the application.

After evaluating the IND and conferring with the other reviewers assigned to the application, the primary reviewer then reconciles the various reviews, and makes a determination on the IND, sometimes in consult with the branch chief. Generally, the primary reviewer presents this decision at a weekly office-wide staff meeting, which provides other reviewers an opportunity to offer input on the IND. The division directors of BEPA, DVRPA, and DARP have final authority over the fate of INDs under their respective divisions.

The IND and the 30-Day IND Review Clock
As it does for drug INDs, the FDA has only 30 days in which to reach a decision on pending biologic INDs. CBER staffers report that difficulties in completing IND reviews within the 30-day review period arise fairly infrequently.

As the 30-day IND review deadline approaches, CBER has two primary options: (1) allow the sponsor to initiate the clinical trials specified in the IND; or (2) issue a clinical hold, the mechanism through which the FDA can delay the clinical trials until pending problems and/or questions are resolved. Through an unofficial procedure called an "informal clinical hold," the center may also ask the sponsor to voluntarily delay the initiation of its clinical trials until CBER can complete its review (for a complete discussion on clinical holds, see below).

If CBER finds no problems with the IND submission, product safety, or the proposed clinical trials, the center "passively" approves the IND by allowing the 30-day review clock to expire without issuing any communication to the sponsor. In this way, INDs are never formally approved by the FDA. Although not required, sponsors are often advised to contact CBER before initiating trials.

According to CBER estimates through 1991, the center passively approves roughly 80 percent of original IND submissions. This percentage has been fairly stable during the past several years, the center reports.

The Clinical Hold

When CBER discovers serious deficiencies that cannot be addressed before or during the IND review process, the center will contact the sponsor within the 30-day review period to delay the clinical trial. The clinical hold is the mechanism that CBER uses to accomplish this.

Through this order, the agency may either delay the initiation of an early-phase trial on the basis of information submitted in the IND, or discontinue an ongoing study based on either a re-review of the original IND or a review of newly submitted clinical protocols, safety reports, protocol amendments, or other information. When a clinical hold is issued, a sponsor must delay or discontinue any trial identified in the order (other studies of the drug, if there are any, may be initiated or continued). The sponsor must address the issue that is the basis for the hold before the order is removed.

The FDA's authority regarding clinical holds is outlined in the IND regulations of 1987. The regulations specify the clinical hold criteria that the FDA applies to the various phases of clinical testing.

Clinical Holds and Phase 1 Trials   One of the principal goals of the 1987 IND regulations was to give sponsors "greater freedom" during the initial stages of clinical research. Therefore, the regulation states that the FDA should not place a clinical hold on a Phase 1 study "unless it presents an unreasonable and significant risk to test subjects." In the regulation's preamble, the FDA states that it will "defer to sponsors on matters of Phase 1 study design," and will not consider a Phase 1 trial's scientific merit in deciding whether it should be allowed to proceed.

The regulation specifies four situations in which the FDA can either delay a Phase 1 study proposed in an IND or discontinue an ongoing Phase 1 trial:

- if human subjects are or would be exposed to an unreasonable and significant risk of illness or injury;

- if the clinical investigators named in the IND are not qualified by reason of their scientific training and experience to conduct the investigation described in the IND;

- if the investigator's brochure (i.e., material supplying drug-related safety and effectiveness information to clinical investigators) is misleading, erroneous, or materially incomplete; or

- if the IND does not contain sufficient information as required under federal regulations to assess the risks that the proposed studies present to subjects.

According to the estimates of CBER staffers, approximately 20 percent of Phase 1 protocols are placed on clinical holds at the IND review stage, although the frequency is likely to vary greatly between product types. Many Phase 1 holds can be lifted relatively quickly, unless additional preclinical testing or product characterization is needed.

Clinical Holds and Phase 2 and 3 Studies   The FDA has greater discretionary powers to delay and discontinue Phase 2 or Phase 3 trials. Current regulations allow the agency to place a clinical hold on a Phase 2 or 3 trial if: (1) any of the four Phase 1 clinical hold criteria outlined above are met; or (2) the "plan or protocol for the investigation is clearly deficient in design to meet its stated objectives."

The second criterion has been somewhat controversial. The FDA had never before held the authority to stop a trial due to design or scientific deficiencies unrelated to patient safety. Some industry officials, particularly those at more experienced firms, believe that companies should be allowed to experiment and establish their own scientific and medical cases, and that FDA involvement should be limited to issues pertaining to patient safety.

In response to questions about its authorities regarding Phase 2 and 3 trials, the FDA stated in the regulation's preamble that the agency "... has the authority and responsibility to establish conditions, including a review of study design, to ensure that a study that is conducted to develop evidence of a [product's] safety and effectiveness is designed to achieve its objectives. Review of the study design may prevent unnecessary mistakes, may assure the adequacy of a study, and may otherwise increase the likelihood that completion of the study will generate the kind of data needed to make a final determination about the [product's] safety and effectiveness."

The agency has indicated, however, that clinical holds are reserved for the most serious design defects. Current regulations place the burden on the FDA to show that a design defect is critical with respect to the purpose of the study.

How Clinical Holds Work   The FDA acknowledges that the imposition of a clinical hold is a relatively informal and flexible process. Given the nature of product development, the agency has resisted suggestions that it formalize the clinical hold process.

Current regulations state that the process will, in many cases, begin with an FDA-sponsor discussion: "Whenever FDA concludes that a deficiency exists in a clinical investigation that may be grounds for the imposition of a clinical hold, FDA will, unless patients are exposed to immediate and serious risk, attempt to discuss and satisfactorily resolve the matter with the sponsor before issuing the clinical hold order."

In certain situations, CBER may ask sponsors, on an informal basis, to voluntarily agree to an extension of the 30-day review to avoid a clinical hold order. This procedure, however, may be much less common in CBER than in CDER. If an IND is mistakenly submitted to another FDA review group and much of the 30-day review clock is consumed while the application is channelled to CBER, the center may ask the sponsor to voluntarily agree to delay its clinical studies so a complete review may be done. Called an "informal clinical hold," this mechanism allows the FDA and the sponsor alike to avoid complications associated with formal clinical hold orders (e.g., paperwork, etc.). In such cases, CBER may set a date by when the company may proceed if not contacted by the center, or promise to contact the sponsor as soon as a decision is reached.

Federal regulations make clear, however, that the FDA is not obligated to initiate a dialogue or to pursue alternate means before issuing a hold: "While the agency is committed to making a good faith attempt to discuss and satisfactorily resolve deficiencies in an IND before considering the need to impose a clinical hold, it does not believe that it is obligated to establish procedural ... [requirements obligating the agency to take such action.] The nature of the agency contact with sponsors will depend on the imminence of hazard to human subjects, on the availability of key agency and sponsor personnel, and on a variety of other factors. For similar reasons, FDA believes that it cannot, in the abstract, specify the extent of notice that can approximately be given before making a hold effective."

CBER may issue a clinical hold order by letter, telephone, or some other method of rapid communication. This order "will identify the studies under the IND to which the hold applies, and will briefly explain the basis for the action." A full written explanation from CBER will follow as "soon as possible, and no more than 30 days after imposition of the clinical hold." The sponsor may appeal the FDA's decision.

When the hold order is issued, identified studies must be delayed or discontinued immediately. If the study has not yet begun, no subjects may be administered the investigational biologic. Ongoing studies placed on clinical holds must be discontinued immediately, and no new subjects may be recruited to the study or placed on the treatment. CBER may, however, permit subjects already on the treatment to continue receiving the experimental biologic.

Under the terms of some clinical hold orders, sponsors may be allowed to begin or to continue the affected investigations when the required modification is instituted and without prior notification of CBER. In all other cases, the investigation may proceed only after the sponsor has notified CBER, and after the company has received the center's authorization to begin or resume the clinical program.

At various times, the FDA has been concerned about the consistency of clinical hold decisions. In mid-1993, for example, CDER established a permanent committee to periodically review clinical hold decisions on drug INDs. Since a top CBER official sits on the committee, panel deliberations could influence CBER's clinical hold policies and practices.

IND Status

CBER uses a fairly sophisticated nomenclature for classifying and tracking the status of INDs under its purview. Although not all elements of the system are relevant to product sponsors, the system is based largely on federal regulations, which specify five status categories for INDs:

Active Status: Generally, an active IND is one under which clinical investigations are being conducted. In other words, the FDA has decided not to delay or suspend the clinical studies proposed in an IND or subsequent protocol amendments. An IND, however, may remain on active status for extended periods even though no trials are being conducted under the application. In such cases, the sponsor may resume clinical studies under the IND without further notification to CBER. In practice, CBER staffers refer to active INDs as "effective" INDs, and use the term "active" in a broader sense for administrative tracking purposes.

Inactive Status: An inactive IND is one under which clinical investigations are not being conducted. There are two ways in which an IND can be placed on inactive status. First, the IND sponsor may ask the FDA to do so, thereby eliminating the sponsor's IND updating and submission requirements. Also, CBER may place the application on inactive status if the agency finds either: (1) that no subjects are enrolled in an IND's clinical studies for a period of two years or more; or (2) that all investigations under an application remain on clinical hold for one year or more.

Clinical Hold. As discussed, a clinical hold is an FDA order to delay a proposed investigation or to suspend an ongoing investigation. If all investigations covered by an IND remain on clinical hold for one year or more, the FDA may place the IND on inactive status. Within CBER, a "partial" clinical hold refers to situations in which one or more of an IND's studies are placed on clinical hold (e.g., when a single investigator is considered unqualified), while the remaining studies are allowed to proceed.

Withdrawn Status. The sponsor of an IND can withdraw an IND at any time and for any reason. When the sponsor does so, all clinical investigations under the IND must be suspended, all investigators notified, and all stocks of the drug returned to the sponsor or otherwise disposed of at the sponsor's request.

IND Termination. The FDA will seek to terminate an IND if the agency is unable to resolve deficiencies in an IND or in the conduct of an investigation through a clinical hold or through less formal channels. Excluding cases in which continuing an investigation will present an immediate danger to clinical subjects, the FDA will issue a proposal to terminate, and offer the sponsor an opportunity to respond before finalizing a termination.

Chapter 6

The Clinical Testing of Biologics

From animal testing through the development of the IND, virtually all preclinical work is undertaken to obtain the FDA's permission to study a new biologic in clinical trials. As the ultimate testing ground for unapproved products, clinical trials produce data that serve as the most critical factor in the FDA's evaluation of a new biologic.

The diversity of biological products makes developing an overview of clinical studies for such products a particularly difficult challenge. In fact, since clinical testing programs are influenced so fundamentally by the nature of the products under study, this diversity generally limits the value of such overviews.

Therefore, this chapter is divided into two sections to highlight some of the important differences in the clinical trial approaches to various biological products. Authored by Biogen, Inc.'s Neil Kirby, Ph.D., the first section addresses the clinical testing of therapeutic products. The second section, on the clinical testing of preventive vaccines, was co-authored by Karen L. Goldenthal, M.D., the acting director of CBER's Division of Vaccine and Related Products Applications, and Loris D. McVittie, Ph.D., acting chief of CBER's Viral Vaccines Branch.

The Clinical Testing of Therapeutic Biological Products

by Neil Kirby, Ph.D.
Manager, Regulatory Affairs
Biogen, Inc.

The objectives of clinical studies with biological products are identical to those for traditional, small-molecule drugs—that is, to demonstrate the product's safety and effectiveness. A clinical trial is a systematic study of a medicinal product in human patients (or in non-patient volunteers), and is designed to identify or verify the investigational product's beneficial and adverse effects (i.e., define the product's risk/benefit relationship). Clinical studies may also be performed to investigate a product's absorption, distribution, and metabolism in healthy volunteers.

A clinical study's design, scope, and conduct depend on several factors, including the type of product being tested, the stage of clinical development, the results of previous testing on the biologic, and the intended clinical application. For example, the design of a clinical trial to study a replacement therapy, such as insulin or human growth hormone, is considerably different from the design of a study involving a pharmacologic therapy, such as the interferons or tissue plasminogen activators.

With this in mind, the following section profiles the typical progression of a therapeutic biologic's clinical development. This basic overview includes specific—though not exhaustive —discussions of clinical trial design issues relating to various biological therapeutic products, including monoclonal antibodies, recombinant proteins, and somatic cell and gene cell therapies.

The Phases of Clinical Investigation

A biologic's clinical development generally proceeds in distinct steps or phases, with each phase building upon information obtained from the previous phase(s). Although there are no statutory requirements mandating a structure or design for clinical studies, federal regulations include a description of the clinical phases (21 CFR 312.21) that provides useful insights into the development process.

However, clinical development programs for new therapeutic agents do not follow a rigidly defined progression. Rather, the nature, focus, and course of the product development process are based on specific product characteristics and the disease being treated. This is particularly relevant for biological therapeutics, which often have complex and pleiotropic effects *in vivo*.

Still, there exists a general framework for all clinical development. It consists of several loosely defined phases that, due largely to wide acceptance by the FDA and medical and scientific research communities, represent *de facto* standards. According to FDA regulations, these phases "describe the usual process of drug development, but they are not statutory requirements. The basis for marketing approval is the adequacy of data available; progression through the particular phases is simply the usual means the sponsor uses to collect the data needed for approval."

Pilot and Phase 1 Clinical Studies   Clinical development often begins with pilot studies, which are designed to explore the product's clinical safety and activity before more formal Phase 1 or combination Phase 1/2 studies are undertaken. Generally, pilot studies are performed in a small number (five to ten) of patients, and often are used to examine products with similar characteristics—for example, to screen a panel of monoclonal antibodies against the same antigen to determine which antibodies deserve further testing. Such studies accelerate the process of identifying products for further testing, and focus resources on products showing early signs of clinical activity.

   In the absence of pilot studies, Phase 1 investigations represent the first human exposure to a therapeutic product. These studies are used to determine the biologic's metabolism and pharmacologic actions, as well as the side effects associated with the product.

   Typically, Phase 1 tests are closely monitored studies involving healthy male volunteers. Because of this intensive in-patient monitoring, several dedicated, specialized Phase 1 clinical units have been established around the world.

   The absence of specific animal reproduction data early in the product development process accounts for the use of male volunteers in such studies. However, in an April 5, 1993 talk paper, the FDA announced plans to issue new guidelines that will encourage sponsors to include more women of child-bearing age in early clinical studies.

   Healthy volunteers represent a homogeneous patient population in which to study a product's safety and pharmacodynamics. Since Phase 1 studies are usually small (20 to 80 subjects), only the most commonly noted adverse effects will likely be identified (e.g., flu-like symptoms after the administration of the interferons). A more detailed analysis of the safety profile is possible only after studying the product in larger patient groups used during the later phases of clinical investigation.

   Phase 1 studies are often used to establish the maximum tolerated dose (MTD) of a product following single and multiple doses. In addition, these

Biologics Development: A Regulatory Overview

studies often provide useful information on the product's pharmacokinetics, absorption, distribution, and metabolism.

Under certain circumstances, such as early-phase testing for oncology products, Phase 1 studies are conducted in seriously ill patients. This is common when a product's mechanism of action is likely to produce toxic effects.

For example, the use of healthy volunteers may be inappropriate in clinical investigations of murine monoclonal antibodies, which are likely to cause the human anti-mouse antibody (HAMA) reaction. If antibodies are raised against the constant region of the injected murine monoclonal antibody, these antibodies are likely to cross-react with a wide range of other monoclonal antibodies of the same species. This would sensitize the volunteer to subsequent administration of a murine antibody. In addition to posing a clinical threat, the HAMA response may affect the pharmacokinetics of the injected monoclonal antibody (i.e., by increasing clearance via complex formation in the liver).

Whenever possible, the new therapeutic's initial dose (whether administered during Phase 1 or pilot studies) should be based on data generated in animal studies. Given the exquisite species specificity often associated with biological products, however, this approach is frequently unavailable. The lack of an appropriate (i.e., responsive) animal model necessitates the use of much lower doses in early phases of clinical investigation. These initial doses are then increased in a cautious, step-wise manner.

Given the biologic response-modifying (BRM) nature of the interferons, Phase 1 studies for these products required novel trial designs. Interferon alpha, for example, was originally studied for oncology indications in which the classical cytotoxic model of drug activity is usually expressed as the MTD. However, the optimal BRM dose is not usually the MTD, since many BRM drugs have a "bell-shaped" or other non-linear dose-response relationship. Therefore, the potential for non-linear dose-response relationships must be considered in designing clinical trials for such products.

Although measuring effectiveness is not typically a primary goal of Phase 1 studies, initial-phase testing in cancer patients should include a basic measure of effectiveness to elicit an early indication of activity. In addition, novel study designs may allow researchers to examine the pharmacodynamic profile of certain biological products. In a 1962 Phase 1 study, for example, interferon alpha demonstrated its antiviral effects in healthy volunteers by inhibiting successful immunization after smallpox vaccination (1). Similarly, the antiviral effects of intranasal interferon alpha were demonstrated during Phase 1 studies in which healthy volunteers receiving the product were challenged with partially attenuated influenza strains and rhinovirus 1B (2).

In addition to their utility in assessing drug safety and tolerability, pilot and Phase 1 studies represent an excellent opportunity to test and validate bioassays to be used in future clinical studies.

Another clinical development approach involves the conduct of combination Phase 1/2 studies. Such tests involve patients with the targeted condition, and are used to explore both safety and effectiveness. These studies tend to employ open-label, dose-escalation designs, and often are performed when the product has been used for other indications or is a second-generation product. Phase 1/2 studies' primary goal is to assess a new product's gross toxicity and to define a therapeutic dose range in the disease population.

Phase 2 Clinical Trials   Generally, Phase 2 studies represent the first controlled clinical trials designed to measure a product's effectiveness in its intended use(s). This phase also presents an important opportunity to evaluate appropriate study endpoints for later trials.

Early Phase 2 clinical studies frequently employ open-label designs. As researchers gain experience with the product, however, placebo controls are often integrated into subsequent studies.

Often divided into Phase 2a (early) and Phase 2b (later), Phase 2 studies are usually conducted in a limited number of patients (100 to 300), and are used to establish an effective dose and regimen for further study in larger clinical trials. As mentioned previously, establishing appropriate doses and regimens is especially relevant for biological response modifiers, which often have non-linear dose-response relationships. During this phase, researchers should also investigate the duration of biological effects, since a prolonged time to response may be relevant in the design of later clinical studies.

The completion of Phase 2 studies represents an important milestone, since the sponsor is poised to begin the larger, Phase 3 trials pivotal to product approval. For this reason, biologic sponsors are often advised to meet with CBER in an "end-of-Phase 2" meeting to discuss clinical plans. According to FDA regulations, the purpose of end-of-Phase 2 meetings "is to determine the safety of proceeding to Phase 3, to evaluate the Phase 3 plan and protocols, and to identify any additional information necessary to support a marketing application for the uses under investigation." As such, the meeting allows the FDA to make recommendations on the design of pivotal clinical studies.

Phase 3 Clinical Trials   The primary objective of a Phase 3 clinical study is to establish a new therapeutic product's safety and effectiveness in a specific patient population. These studies often involve large numbers of patients (200 to 1,000 or more), and are conducted under conditions more closely resembling those under which the drug is to be marketed. Therefore, Phase 3 trials not only provide the key evidence of the biologic's safety and effectiveness, but important insights for product labeling as well.

Given their importance to the FDA's approval decision, these pivotal studies must meet particularly high scientific standards. Phase 3 studies tend to be randomized, controlled, double-blind, multicenter studies (although other designs also may be acceptable), and are designed to provide statistical proof of effectiveness and to further define drug-related adverse events (for a further discussion of these standards, see discussion on clinical trial design below). Usually, these studies constitute the "adequate and well-controlled" clinical investigations necessary to support product licensure (i.e., studies pivotal to the approval of a product).

Exclusion criteria deemed prudent for earlier clinical studies are often liberalized somewhat during Phase 3 clinical investigations. In short, Phase 3 study populations should more closely resemble the population likely to use the product if approved. Specific patient groups not treated in Phase 3 studies (and therefore not included in the labeling for the product) are often studied in Phase 4, or postmarketing, clinical studies (see below).

In approving new drugs, the Center for Drug Evaluation and Research (CDER) typically interprets statute and good science as requiring at least two "adequate and well-controlled" studies to establish product effectiveness. Although CDER has approved drugs on the basis of only one, compelling pivotal study, such situations are generally restricted to therapies for AIDS and other life-threatening conditions.

In the past, CBER has based its approval of certain biological products on a single "adequate and well-controlled" study. Since each new therapeutic product is assessed on a case-by-case basis, it is important to note that there exist no criteria for determining whether a product will need one or two "adequate and well-controlled" studies to support licensure. Also, while product approvals are always based on the scientific merit of available evidence, it is clear that the FDA wants to avoid having two evidentiary standards (i.e., one for biological products and another for conventional drugs).

Because of these apparently differing standards, joint CDER-CBER reviews pose an interesting challenge when designing a clinical development program. The Peripheral and Central Nervous System Drug Advisory Committee—a CDER committee—recently considered a biological product, Betaseron® (recombinant interferon beta-1b), for the treatment of multiple sclerosis. This was the first biological product discussed by the committee, which is accustomed to making recommendations based on two pivotal studies. During deliberations, several members of the committee wanted the FDA to inform sponsors that pursuing approval with only one pivotal clinical study is a "dangerous strategy." Despite such concerns, however, the committee did recommend approval for this product. Because there will be other joint reviews, sponsors should consider the potential for intercenter evaluation when planning clinical development programs.

CBER's product-by-process approach to biologics (i.e., the characteristics of a product are critically dependent upon the manufacturing method) has important implications for clinical trials. Most importantly, the investigational product administered during Phase 3 pivotal studies should be manufactured through the process to be used for commercial production.

Phase 3 pivotal clinical trials often present a good opportunity to examine a product's economic profile (i.e., cost vs. benefit), which is especially relevant given growing concerns about health-care costs. Integrated clinical/economic studies are possible, with the nature of the biologic dictating the variables to be assessed (for example, reduction in hospitalization time). However, it is important that any economic analysis not impede the primary intent of the Phase 3 clinical study—economic analysis is often important in achieving marketing success, but is not an FDA requirement. Economic analyses may be particularly relevant to biotechnology-derived products, which often have higher prices due to high manufacturing costs.

It is worth reiterating that the phases of clinical study described above are merely examples of the types of studies performed in the past. It is unlikely that a sponsor will use all phases (i.e., pilot, Phase 1, Phase 1/2, Phase 2, and Phase 3) in such a sequential manner during a clinical development program.

To a great extent, a new biologic's clinical development program will be a function of the product's nature and the disease to be treated.

Phase 4 Clinical Trials   Phase 4 trials are postlicensure studies designed to confirm the safety of a therapeutic product in large populations. Such studies are frequently required as a condition of approval (see below), and are, therefore, designed to confirm the results observed in earlier clinical trials.

Phase 4 clinical studies are useful in observing drug interactions that may occur with concomitantly administered drugs in the general population, and are often designed to study licensed products' safety and effectiveness for unlicensed indications or to identify a competitive advantage over similar agents (e.g., less antibody production). In addition, Phase 4 studies often are performed in patient subpopulations not previously studied in other trials (e.g. children and the elderly).

In general, Phase 4 clinical trials tend to be less well-controlled than Phase 3 studies. However, studies designed to support the approval of new indications should meet the standards applied to pivotal trials.

Issues in Clinical Trial Design
Clinical trial design is critical to the scientific integrity of pivotal studies and to the credibility of the data produced. The following discussion profiles several of the most important clinical trial design issues for therapeutic biological products—controls, sample size, and clinical endpoints—and related issues, such as risk/benefit and clinical trial supplies.

Controls   Many clinical trials are comparative in nature—in other words, the study compares a group of patients receiving a new treatment with another group (i.e., a control group) receiving some other form of therapy. In federal regulations, the FDA identifies five types of controls (21 CFR 314.126): (1) placebo controls; (2) active concurrent controls; (3) dose comparison controls; (4) no treatment concurrent controls; and (5) historical controls.

Although the FDA and the medical community view placebo-controlled studies as the "gold standard," the type of control employed in a clinical trial depends on the product, the indication, and the phase of investigation. Active concurrent controls (i.e., an already approved therapy for the same indication) are usually appropriate for studies in which the use of a placebo would be unethical—for example, in the treatment of most malignancies for which there are approved therapies. Dose comparison-controlled studies utilize two or more doses of a new therapeutic product, and are useful in evaluating dose-response relationships. No treatment concurrent control studies are appropriate when objective measurements of effectiveness are available and placebo effects are negligible.

Historical controls can be used in certain trials, such as those involving diseases with high mortality rates. For example, the pivotal studies of Intron® A (recombinant interferon alpha-2) as a treatment for hairy cell leukemia utilized a historical control group to compare survival rates of treated and untreated patients. This study demonstrated a superior actuarial survival in Intron® A patients compared with a historical control group of untreated patients.

The choice of the control often has a major effect on the size of a clinical study. Studies to investigate the equivalence of two therapies will likely be sized quite differently than a similar study to identify a difference between the two therapies (depending upon the size of difference expected). In general, trials involving significant differences in therapeutic effects, which are easier to detect, require fewer patients.

Sample Size   Determining the number of patients needed for a trial is a fundamental design issue for pivotal studies. Therefore, statisticians should be intimately involved in the design of clinical studies to determine the sample size. In general, the sample size must be sufficient to provide adequate "power"—the probability that a specific treatment difference will be detected if it exists.

Statistical power is a direct function of the number of patients studied (the more patients studied, the greater the power), which is, in turn, influenced by the magnitude of the differences between the comparative therapies and the variability of the data. In general, sponsors should aim for at least 80 percent power when planning a clinical study, although this may not be appropriate in all situations.

The statistical methods used to calculate sample size are based on the primary measure of outcome, the method of analysis, and the anticipated results of standard therapy. For any sample-size analysis, sponsors should make provisions to compensate for expected study drop-outs, protocol deviations, and other possible developments, so that the study enrolls the appropriate number of evaluable patients.

Sponsors must view statistical issues as an integral part of study design. When such issues are overlooked, statisticians are often expected to analyze the data creatively to compensate for a poorly designed study—an unenviable situation.

Statistical reasoning is not the only factor that affects sample size. Limited patient populations for certain orphan indications often make large pivotal clinical trials impossible. For example, the approval of agluceredase (Ceredase®), a treatment for Gaucher's disease (a rare disease caused by deficient glucocerebrosidase), was based on the treatment of only 31 patients.

Clinical data analyses conducted during human trials also affect sample size. The review of accumulating data (i.e., interim analysis) poses statistical, medical, and ethical issues, and requires that a trial enroll enough patients to offset the effects of repeated significance testing (i.e., generally, the more frequent the interim analyses, the larger the population needed). Therefore, the protocol must specify any interim analysis decision rules prior to a trial's initiation.

External data-monitoring committees are often used to evaluate the results of interim analyses and to make recommendations to the sponsor. The protocol should also clearly outline the responsibilities of such committees.

Clinical Endpoints   Clinical endpoints are generally clinical events (e.g., death, loss of vision) or measurements (e.g., blood pressure) used to assess a therapy's effectiveness. In the context of a clinical trial, the assessment of a biologic's effectiveness focuses on the product's ability to prevent a harmful clinical event such as death or illness, or to otherwise modify a clinical endpoint in a manner that has clear clinical benefits for the patient.

Primary endpoints are generally clinical events (such as death or myocardial infarction), and should be chosen because the absence or presence of the event results in a clear and definite clinical effect on the patient (i.e., it is clinically relevant). In general, studies of life-threatening illnesses should include an evaluation of the investigational product's effect on mortality. Study endpoints should be as objective as possible, and should utilize validated methodologies whenever available.

The use of surrogate endpoints, or surrogate markers, in clinical studies has sparked considerable debate. A surrogate endpoint is an event or a measurement correlated with, and a likely predictor of, an effect on a primary (clinically relevant) endpoint. One scientifically accepted surrogate endpoint is the level of cholesterol in patients, which has been defined as an appropriate predictor of eventual coronary or cerebral artery disease. In contrast, the CD4 cell count—an indication of the body's response to AIDS—alone has not, to this point, been accepted as an appropriate surrogate endpoint in measuring the effectiveness of HIV therapies.

CBER has accepted laboratory measurements as appropriate surrogate endpoints in approving some biological products. For example, the Biological Response Modifiers Advisory Committee agreed that, in Phase 3 clinical studies of Intron A® as a treatment for chronic hepatitis B, the virus markers HBeAg and HBV-DNA were valid and important endpoints.

Sponsors should recognize that not all surrogate endpoints are clinically relevant in the eyes of FDA reviewers and medical researchers. Some endpoints, in fact, may be relevant only for products with particular mechanisms of action. Therefore, sponsors must rigorously defend the use of unestablished surrogate endpoints, particularly in pivotal Phase 3 studies.

A January 1993 regulation providing for the accelerated approval of drugs and biologics intended to treat serious or life-threatening diseases includes a mechanism for product approval based on surrogate endpoints. In such cases, the surrogate endpoint must be reasonably likely to predict a clinical benefit, and the sponsor must support the choice of the proposed surrogate. In addition, the regulation also permits approvals based on a clinical endpoint other than survival or irreversible morbidity, pending completion of studies to establish and define the degree of the clinical benefit. When approval is based on surrogate endpoints, or on the effect on a clinical endpoint other than survival or irreversible morbidity, the sponsor must conduct postmarketing clinical studies to verify and confirm the product's clinical benefit and to resolve any remaining uncertainties regarding the surrogate marker's clinical relevance. These postmarketing studies must meet the general criteria for "adequate and well-controlled" trials.

Risk/Benefit Relationship   The risk/benefit relationship, or ratio, refers to the potential benefits a patient might derive from a product compared to the likely risks incurred through the product's use. A biologic with a favorable risk/benefit ratio offers a significant potential benefit while presenting few potential risks (i.e., side effects).

The acceptance of new therapeutic products with unfavorable risk/benefit ratios (i.e., high potential for adverse effects) often varies among patients. For example, one cancer patient may accept a potentially life-saving therapy with a severe toxic profile, since it may represent the only chance for a cure or a remission, whereas another patient may not be willing to accept the same risks because of the expected reduction in quality of life.

High-dose interleukin-2 (IL-2) is an example of a biological product with a relatively unfavorable risk/benefit ratio. Although the product is associated with considerable toxicity (e.g., capillary leakage syndrome), IL-2 (either alone or with lymphocyte-activated killer cells) produces significant tumor regression in melanoma and renal cell carcinoma and is an accepted therapy for certain patients.

Effectiveness and safety are not absolutes; each must be evaluated in the context of the disease being treated. Evaluating a new therapeutic product's risk/benefit relationship is an important part of the development and approval process, and is, therefore, an integral part of clinical trial design.

Clinical Trial Supplies   The production and ultimate quality of materials used in clinical studies have particular relevance for biological products, which traditionally have been defined by the process of manufacture (i.e., product-by-process). All material used in clinical studies must be manufactured in accordance with current Good Manufacturing Practice (cGMP) in a capable facility using qualified equipment and validated processes (see Chapter 12).

It is also critical that sponsors thoroughly assess and evaluate the impact of manufacturing and formulation changes made during the development program. Changes to the purification process may affect viral clearance and/or inactivation parameters, and may pose potential safety problems in clinical use.

Similarly, purification changes may affect the glycosylation profile of proteins expressed by mammalian cells. These glycosylation changes may have profound effects on both the pharmacodynamic and pharmacokinetic profiles of biological products in the clinic.

It is generally understood that manufacturing changes are necessary during a new product's development. Whenever possible, however, sponsors should anticipate these changes, and discuss them with the FDA prior to their implementation. It is worth reiterating that Phase 3 pivotal clinical studies should be performed with material manufactured and formulated according to the process intended for commercialization and described in the PLA and ELA.

Specific Issues in the Design of Clinical Studies of Biotechnology-Derived Therapeutic Products

In addition to the broad-based clinical trial issues discussed above, there are many more that are specific to each class of biological therapeutic product. The following discussion highlights some of the specific issues relating to the design and monitoring of clinical studies for therapeutic products manufactured through biotechnology processes (i.e., monoclonal antibodies, recombinant proteins, and somatic cell and gene therapies).

Monoclonal Antibodies Currently, there is a great deal of discussion regarding the risks posed by murine xenotropic retroviruses in monoclonal antibody preparations and the need for their removal prior to pilot or Phase 1 clinical studies. Since validating the removal and inactivation of viral contaminants significantly delays the development of monoclonal antibody therapies, some have proposed that such testing is unnecessary prior to early pilot clinical studies in limited numbers of terminally ill patients.

Eliminating such validation studies would give patients earlier access to more monoclonal antibody treatments for cancer. As it does for most biotechnology-derived products, the FDA advocates a case-by-case, flexible approach to the development of monoclonal antibodies.

Risk assessment in early pilot or Phase 1 testing of monoclonal antibodies must include an analysis of any cross-reactivity seen *in vitro*. In general, the more specific the antibody *in vitro*, the less chance that the antibody will produce adverse events *in vivo*.

Sponsors must examine the degree of cross-reactivity, the dose, and the amount of preclinical data before initiating clinical testing. The FDA does recognize that some antigens are very specific to humans, and that appropriate animal models may not be available. Therefore, in the absence of appropriate preclinical data on monoclonal antibodies due to species specificity, initial clinical studies must involve extremely low doses.

Researchers anticipate that the advent of humanized and genetically engineered monoclonal antibodies will resolve many of the problems encountered with xenogenic monoclonal antibody products (including the HAMA response).

Recombinant DNA-derived Proteins   During the clinical testing of recombinant DNA-derived proteins, it is critical that the sponsor closely monitor the production of antibodies to the injected protein. Although the production of antibodies to natural-sequence human proteins is unlikely, it is important to monitor for such antibodies, which may interfere with the therapeutic product's activity.

The presence of neutralizing antibodies will not always result in a loss of effectiveness, however. For example, approximately 45 percent of multiple sclerosis patients treated with interferon beta-1b eventually developed neutralizing antibodies during a clinical study. Apparently, the presence of these antibodies did not affect the product's safety or effectiveness, although it is possible that the product exhibited effects before the emergence of antibodies.

In addition to performing analyses for product-specific antibodies, sponsors may want to monitor the development of antibodies against host organism proteins, residual amounts of which may be present in the product. Clinical studies on a recombinant protein produced in *E. coli*, for example, should include testing for antibodies against *E. coli* proteins, which may be present in the finished product at levels undetectable through routine quality control release testing. Excluding patients with known hypersensitivity to *E. coli*-derived proteins from studies on such products might be appropriate as well.

Similarly, sponsors should exclude patients with a known sensitivity to beta-lactam antibiotics from clinical studies of products manufactured in the presence of such antibiotics (e.g. penicillin). For this reason, the FDA has discouraged the use of these antibiotics in the production of recombinant therapeutic proteins.

Somatic Cell Therapy and Gene Therapy   The use of somatic cell and gene therapies to treat serious, life-threatening diseases is an emerging area of research. As such, the design of clinical studies for these therapeutic modalities presents several important, cutting-edge issues.

114

Somatic cell therapy involves the administration of autologous, allogenic or xenogenic living cells, which have been manipulated or processed *ex vivo*. The infusion of lymphoid cells, such as lymphocyte activated killer (LAK) cells, is an example of such therapies.

In designing clinical studies for somatic cell therapies, sponsors should give special consideration to several issues, including:

• the survival of the infused cells;
• the localization of the infused cells;
• the quantitation of the products made by the cells; and
• the replication of any viral vectors *in vivo*.

As in any clinical investigation, the analysis of adverse events is an important part of such studies. Sponsors should carefully monitor the presence of infections that might be related to the infused cells. In addition, the use of certain viral vectors will necessitate the testing of patient contacts (including health-care providers) to confirm the lack of infectious spread. Depending upon the nature of the infused cells, patient follow-up may be necessary for the recipients' remaining lifetimes. Since it is unlikely that the infused cells' effects on fetal development will be understood fully, sponsors should carefully assess the use of somatic cell therapy in pregnant women.

One of the most exciting recent medical developments is the ability to modify the genetic material of living cells (either *ex vivo* or *in vivo*) for therapeutic purposes, a process known as gene therapy. Gene therapy would include, for example, the transfer of part of the normal human gene to the respiratory epithelium of patients with cystic fibrosis.

Some special considerations in the design of clinical studies for this potentially powerful therapeutic tool include the following:

• the product of the inserted gene must be studied (it is possible that this product could be antigenic); and

• patients must be monitored for immune responses to the "carrier" cells, graft-versus-host disease responses and their clinical sequelae, autoimmune responses during therapy, and allergic responses to therapy.

Given the nature of gene therapy, it is likely that such studies will not be performed in healthy volunteers.

A more detailed discussion of relevant issues in the design and conduct of clinical studies for somatic cell therapy and gene therapy can be found in a document entitled *Points to Consider in Human Somatic Cell Therapy and Gene Therapy* (August 1991). In addition, CBER recently established the Division of Cellular and Gene Therapies within the Office of Therapeutics Research and Review to address issues relating to these new therapies.

Clinical Testing and Serious or Life-Threatening Conditions

Since CBER regulates therapeutic products for cancer and other serious and life-threatening conditions, it is worth discussing the special mechanisms available for the development and approval of such therapies. These mechanisms are designed to expedite the availability of new therapies for life-threatening and seriously debilitating diseases.

The FDA's Subpart E regulations (21 CFR 312 Subpart E) provide for accelerated clinical development and FDA review of promising new therapies for such indications. Under the mechanism detailed in these regulations, the FDA and sponsor meet after Phase 1 studies to discuss designs for expanded Phase 2 studies. Often referred to as Phase 2/3 studies, these trials are designed to permit the FDA to reach an early decision on the product's safety and effectiveness. Since Phase 2/3 studies involve fewer patients than conventional Phase 3 trials, the FDA may seek the sponsor's commitment to conduct postmarketing studies to evaluate the biologic further.

While Subpart E describes a method for expediting biologic development and approval, a vehicle called the treatment IND (or treatment protocol) provides for early access to promising, but as-yet unapproved, investigational therapeutics. The FDA grants treatment INDs and protocols only if: (1) the product is intended to treat a serious or immediately life-threatening disease; (2) there are no comparable or satisfactory therapies to treat that stage of the disease in the intended patient population; (3) the product is under investigation in a controlled clinical trial under an IND in effect for the trial, or all clinical trials have been completed; and (4) the sponsor of the controlled trial is actively pursuing approval of the investigational product with due diligence.

In the early 1990s, the FDA also introduced another program under which desperately needed drugs could be made available to patients before approval. Unlike the treatment IND, the "parallel track" mechanism is limited to products used to treat patients with AIDS and other HIV-related diseases. Parallel track is also different in that it may be initiated when the evidence of a drug's effectiveness is insufficient to meet the threshhold necessary for a treatment IND. Bristol-Myers Squibb initiated the first parallel track program in 1992 for its AIDS drug d4T (stavudine).

Given the availability of such mechanisms, therapeutic products for serious or life-threatening diseases will likely have a different clinical development profile than "conventional" products. Because of this, the FDA will, in most of these cases, be intimately involved in the design of appropriate clinical trials to expedite the availability of such life-saving products.

The basic objectives of clinical studies involving therapeutic biological products are identical to those of studies performed on traditional, small-molecule pharmaceuticals. However, the design and conduct of such studies present certain challenges not generally associated with traditional drugs. As new technologies and novel products find applications in clinical studies, these challenges will surely increase.

Therefore, researchers engaged in "leading edge" technology must adopt a flexible approach to the clinical development of their therapeutic agents. Precedents set through the development of earlier products may not always be applicable as technology and knowledge advance, and sponsors must work closely with the FDA throughout the development process to speed the availability of important new products. Timely and effective dialogue with CBER during the development of new therapeutic biological products is essential.

Editor's note: For a complete listing of CBER guidelines, including those relevant to the clinical testing of biological products, see Appendix 1.

This chapter represents the author's assessment of the requirements for therapeutic clinical trials and does not necessarily represent the views of Biogen, Inc.

References

(1) *Effect of Interferon on Vaccination in Volunteers* (a report to the Medical Research Council from the Scientific Committee on Interferon), Lancet 873-875 (1962).

(2) Phillpotts, R.J., Tyrrell, D.A.J., Shepherd, W.M. and F r e e s t o n e, D.S., *Intranasal Interferon ("Wellferon") Prophylaxis Against Rhinovirus and Influenza Virus in Volunteers*, 4 Journal of Interferon Research 535-541 (1984).

(3) 21 CFR Parts 314 and 601.

The Clinical Testing of Preventive Vaccines

by Karen L. Goldenthal, M.D., and Loris D. McVittie, Ph.D.
Division of Vaccines and Related Products Applications
Office of Vaccine Research and Review
Center for Biologics Evaluation and Research

Vaccines provide one of the most effective means of preventing potentially serious infections and play a critical role in public health management. In addition to the direct protection afforded an immunized individual, transmission of infectious diseases from person to person may be interrupted if enough individuals in the community are vaccinated. The contribution of vaccines toward controlling many of the once common childhood infectious diseases has been particularly striking (1-5).

Manufacturers continue to seek ways to improve current vaccines for existing clinical indications and to develop new immunogens for both pediatric and adult use. There is considerable interest in developing new combination vaccines, often composed of already licensed products, with the goal of simplifying recommended immunization schedules and delivering the same number of antigens with fewer needlesticks (3,6). There has also been much interest in developing less reactogenic component vaccines to replace existing whole cell vaccines (e.g., pertussis and typhoid) (3). Investigations with novel adjuvants are focusing largely on the products' abilities to enhance the immune response of certain relatively weak recombinant or peptide antigens (7). Additional polysaccharide conjugate vaccines are being developed to expand indications to younger age groups (3). Also, new indications for which there are currently no licensed vaccines are being pursued in investigational studies (e.g., HIV vaccines) (8,9). Clearly, the target populations and risk/benefit contexts are quite varied in the examples just presented, and the clinical development program must take each unique situation into account. For all vaccines, however, risk/benefit assessments during both the investigational and licensing stages of necessity must consider the effect on a usually healthy target population.

Federal Regulations Pertaining to Vaccines

In the United States, vaccines are regulated as biological products. To obtain marketing approval for a vaccine, a sponsor must submit both product and establishment licensing applications. In addition to the regulations pertaining to IND studies (see Chapter 4), sponsors will find that pertinent biologics regulations are invaluable when planning clinical studies to support licensure: 21 CFR 600 (Biological products: general), 21 CFR 601 (Licensing), 21 CFR 610 (General biological products standards), 21 CFR 620 (Additional standards for bacterial products), and 21 CFR 630 (Additional standards for viral vaccines). In some instances, the regulations delineate specific minimum requirements for clinical trials to qualify a vaccine for licensure.

Phases of Clinical Development

Phase 1 vaccine trials are primarily designed to evaluate safety and immunogenicity. More than one vaccine dose may be evaluated in a Phase 1 study, usually in a step-wise manner of increasing dose by group. The dose-ranging groups may be studied concurrently or sequentially, depending on the reactogenicity detected in preclinical models or clinical studies of similar products. Usually, clinical evaluation of new vaccines starts with initial Phase 1 testing in healthy adult volunteers who are at low risk of contracting the infection against which the vaccine is indicated.

During the Phase 1 studies, clinical evaluations should include both local injection site reactions and systemic reactions. The prototype case report form (CRF) or subject diary should state each item to be assessed. At predefined intervals, periodic assessments of the local injection site and systemic signs and symptoms will normally be recorded for at least 72 hours post-vaccination. Evaluations of the injection site usually include a quantitative measure of objective items (e.g., erythema and induration) as well as assessments of subjective items, including pain and tenderness. Systemic evaluation includes oral temperatures of vaccinees recorded during this period. Subjective systemic symptoms such as fatigue, myalgias, headache, and nausea should also be recorded.

Subjects who receive live vaccines should be monitored for adverse reactions for a minimum of 14 days post-vaccination or longer as indicated by the particular vaccine. Evaluation of serum chemistry and hematology values may be appropriate in addition to special evaluations that are product-specific (e.g., CD4 cell counts in HIV vaccine recipients). In addition, subjects should be questioned about adverse events when they return for immunogenicity studies at several weeks to one month post-immunization. Because the typical population for vaccine studies consists of healthy subjects, the toxicity criteria for discontinuing further vaccinations should be more restrictive than for many therapeutic settings (e.g., an anti-neoplastic agent in an oncology study).

Environmental analysis statements are required during product development (21 CFR 312.23(a)) either in the form of an environmental assessment or a claim for categorical exclusion. Relevant for most IND studies, claims for categorical exclusion apply when no significant environmental effects are expected (see also 21 CFR 25.24(c)(4)). In INDs for products such as live viral vaccines or vectors, live bacterial vaccines, or parasite-infected mosquitoes used for challenge studies of normal volunteers in vaccine trials, sponsors should describe their proposed procedures for containment of the live biological material during their clinical studies and should provide data on the expected survival of the organism in the environment (if applicable). Shedding of live vaccine organisms should be evaluated in the first Phase 1 study. The isolation of volunteers early in clinical development may be necessary to assess the shedding of attenuated and revertant organisms and the potential for spread to contacts. FDA reviews the sponsor's summary of containment procedures to determine whether such procedures are adequate.

Phase 1 and Phase 2 vaccine studies are not always distinct. In general, however, Phase 2 vaccine studies involve more subjects and are designed to provide more definitive dose and schedule data. At this stage, as applicable, immunogenicity and safety studies are conducted in high-risk populations. Ideally, a preferred formulation, dose, and regimen should be established in a Phase 2 study and then fully evaluated in an efficacy study. Also, concurrent with Phase 1 and Phase 2 studies, background epidemiological data should be obtained on the target population of at-risk individuals. "Breakthrough" infections in vaccinees may be assessed in Phase 2, providing a preliminary suggestion of efficacy in some settings. Laboratory assays that may be used for the case definition for an efficacy trial should be developed no later than Phase 1-2 (e.g., serological tests used to distinguish the immune response in volunteers elicited by vaccination from that elicited by wild-type infection).

The evaluation of interference between the investigational vaccine and already licensed vaccines that would likely be administered concurrently (e.g., during infancy or to travelers) should be considered, preferably early in the clinical development plan. Similar concerns pertain to the concurrent administration of live oral vaccines with oral drugs, and immune globulins with vaccines. Clinical studies should assess the influence of maternal antibodies on the immune response for neonates or older infants if a vaccine will be recommended for individuals in this age range.

Phase 3 studies are larger-scale trials that would, in most cases, include the pivotal efficacy trials as well as expanded safety studies. It is highly recommended that plans for Phase 3 studies be discussed with CBER well in advance of study implementation to ensure that studies are adequately designed to meet intended goals, including support of product licensure.

Exhibit 1

Vaccine Efficacy Protocol Design

The following items should be defined in the efficacy trial protocol:

A. A statement of the study hypothesis.

B. The study population.
- Inclusion/exclusion criteria, including age, sex, health status, laboratory values, concurrent medications, and serological status relevant to the antigen(s) of interest.
- Background data on disease incidence in target population (see below).

C. Vaccine dose, schedule, route, and anatomic site of administration.

D. Study design.
- Control group.
- Randomization scheme/plan for its implementation.
- Study masking.
- Items to be assessed/time schedule: safety, immunogenicity, laboratory parameters, cultures, clinical findings.

- Case definition.
- A clear statement of primary and secondary endpoints to be used to define efficacy.

E. Prototype Case Report Form (CRF).

F. Surveillance Plans.

G. Statistical Section.
- Background data to support assumptions used for statistical calculations.
- Minimal treatment effect detectable with adequate power (especially for the primary endpoint).
- Prospective analysis plan for each endpoint.
- A statistical presentation for the primary efficacy endpoint that considers not only the point estimate of efficacy but also the 95 percent confidence limits.
- Adverse reaction rate comparisons between groups.
- Expected number of drop-outs and losses-to-follow-up.
- Interim analyses (see discussion of Data and Safety Monitoring Board (DSMB) below).
- Justification for trial sample size based on the preceding items in this section.

H. Plan for an independent DSMB.

Vaccine Efficacy Protocols and Trials   Elements of efficacy protocols and the conduct of efficacy trials will be considered in this section. Sponsors should provide a rationale for the vaccine dose and schedule based on the Phase 1-2 studies. The study should be carefully designed to allow a valid estimate of efficacy. Most vaccine efficacy trials have a double-blind, randomized design with one of the following types of controls: a placebo, a licensed vaccine for a non-relevant antigen, or a licensed vaccine for the same indication as the experimental vaccine. The latter control often presents problems due to the large sample size required for a comparative trial. In a trial in which multiple immunizations are necessary to attain optimal immune responses, the possibility that maximum efficacy might not be attained until after the final immunization should be considered in the sample size calculations.

Occasionally, less desirable study designs have been proposed and implemented. In some instances, case-control studies have been accepted, but these pose many potential problems relating to difficulties in ensuring comparability of cases and controls with regard to the likelihood of immunization.

The use of historical rather than concurrent randomized controls has also been considered. However, because of the many possible variations in subject baseline characteristics, changing standards of health care, and altering epidemiology of disease, historical controls are rarely acceptable for establishing vaccine efficacy for licensure purposes.

A vaccine efficacy trial will usually have a single primary endpoint. In most circumstances, this is a laboratory test(s) result, clinical finding or outcome, or a combination thereof. A well-defined and validated case definition of the disease or infection(s) under study plays a critical role in a vaccine efficacy trial (10-12). Prospectively defining a "definite" and "probable" case may be advantageous in some settings. In addition, sometimes it is desirable to not only document the occurrence of a particular clinical infectious disease but also to prospectively plan to quantify its severity. The CRF and subject diary, as applicable, should be submitted with the protocol and should be designed to systematically assess the primary and secondary efficacy endpoints, including safety parameters and the quantification of severity as desired.

Often, it is desirable to observe the duration of efficacy and immunity over a period of several years. Ideally, the correlation of protection with the postvaccination immune response at selected time points (e.g., one month postvaccination and at other specified time points) should be evaluated. This is particularly important if a serological correlate of protection has not been determined previously.

Background epidemiological data, including seroincidence data, should be submitted to the IND to justify the geographic location and study population. Geographic strain specificity should be determined so that the vaccine will include antigen(s) that will be relevant to the infectious organisms occurring in the intended efficacy trial population.

A special mention should be made of vaccine efficacy studies conducted outside of the United States. In some instances, foreign efficacy trials are performed because the disease of interest has a low incidence in the United States. Vaccines for indications such as the prevention of typhoid fever, Japanese encephalitis, and pertussis have been licensed using foreign efficacy data. Immunogenicity differences between populations may result from differences in factors such as genetics, nutritional status, background infections, and prior exposure of vaccinees to the infectious agent; therefore, geographic-specific safety and immunogenicity data should be obtained in the population in which the efficacy trial will be performed (13,14).

Logistical issues such as vaccine distribution and data collection systems should be addressed during the Phase 1-2 studies that precede the efficacy trial. Additionally, safety and immunogenicity data on the use of the vaccine in U. S. populations will be required for licensure in the United States. Any foreign study that is not performed under an IND but that is intended to support licensure must have been conducted by qualified investigators and with adequate protection of human subjects. Additionally, such clinical research must be conducted in accordance with the ethical principles stated in the "Declaration of Helsinki" or the laws and regulations of the country in which the research was conducted, whichever represents the greater protection for study subjects (21 CFR 312.120).

A vaccine efficacy trial should have a formally established independent Data and Safety Monitoring Board (DSMB). DSMB members should have relevant multidisciplinary representation (e.g., statisticians, clinicians with clinical trial experience) and be independent of the sponsor and investigators. If the DSMB will conduct an interim analysis of the data, the following information should be provided prospectively in the protocol: the conditions (e.g., specific schedule, number of events, etc.) which must be met before performing the interim analysis; the criteria for early termination of the study; and a list of the individuals who will have access to the interim data. A detailed statistical plan for any interim analysis should be included with the original submission of the protocol. Any unmasked review of the data, including any review of the data by treatment group (regardless of whether the identity of the treatment is revealed), would be considered an interim analysis. A list of DSMB members and individuals with access to the interim analysis data should be submitted well in advance of such data reviews (15-20).

Additional Licensing Considerations

Evidence of safety and efficacy must be available prior to the licensing of any product. For new vaccines, efficacy is usually assessed in trials demonstrating prevention as discussed above. However, serological endpoints may serve as surrogates for clinical efficacy and, therefore, may be adequate to substantiate effectiveness for licensure when a previously accepted correlation between the surrogate and clinical effectiveness already exists (21 CFR 601.25(d)(2)). Usually, a surrogate for vaccine efficacy would be identified by prior successful clinical efficacy trials with clinical endpoints.

Evaluation of the validity of surrogate serological endpoints for a particular setting often includes assessments of both quantitative antibody titers and functional assays, as well as the physicochemical characteristics of the vaccine. For example, the Vaccines and Related Biological Products Advisory Committee met on September 5, 1991, to consider whether surrogate serological data should be used as the primary evidence of efficacy to support licensure of new *Haemophilus influenzae* type b (Hib) vaccines; criteria were discussed for determining comparability of the immune response elicited by the investigational vaccine to that elicited by licensed Hib conjugate vaccines previously shown to provide protection in clinical efficacy trials (21, 22). At the October 28, 1992, meeting of that committee, a specific Hib conjugate vaccine was recommended for approval and subsequently licensed using serological measures as a primary efficacy parameter.

The use of surrogate serological endpoints is also often considered for multivalent products for a single etiologic agent where an immunological correlate of protection has been determined for one or more of the serotypes. In this setting, it may not be possible to show clinical efficacy for each serotype because of problems such as low incidences of disease caused by certain serotypes. For example, clinical efficacy was demonstrated only for the A and C components of the A-C-Y-W135 meningococcal capsular polysaccharide vaccine prior to licensure; efficacy was inferred for the less common serotypes Y and W135, which each elicited high levels of bactericidal antibodies in volunteers (23). Serological endpoints have also been used to license combination vaccines with components consisting of previously licensed products.

It is recommended that the vaccine product used in pivotal trials be manufactured according to the procedures for which licensure will be sought.

However, occasionally manufacturing changes are made (e.g., in purification scheme or production scale) during the interval between pivotal trial execution and license application submission. Such situations often will require that sponsors make direct comparisons between the test lot(s) used in the pivotal trial and the final lots intended for marketing. Such comparisons are made in "bridging studies," which are designed to demonstrate immunological equivalence between the products and, thus, allow the pivotal trial clinical data to be supportive of final product licensure. Bridging studies, which should be of adequate statistical power and design, are usually based on attaining similar immune responses between concurrent study groups.

In addition, studies that demonstrate the equivalence of different administration schedules or different doses may be required to allow certain clinical data to support product licensure or to support product labeling statements. In some instances, these studies may be performed postlicensure, with the results submitted as a product license amendment (e.g., to expand the range of approved product administration regimens). Protocols for studies with a licensed product that are intended to support a labeling change should be conducted under an IND and should be submitted reasonably in advance of the study's implementation to allow for CBER review.

Licensure decisions also will be affected by the quality and quantity of available safety data. Manufacturers may be asked to perform specific Phase 4 postlicensure studies to provide additional assessments of less common or rare adverse reactions or to evaluate alternative use patterns.

Prior to licensure of a new vaccine or the approval of a significant new indication for a currently licensed vaccine, manufacturers will normally present their data before the Vaccines and Related Biological Products Advisory Committee (VRBPAC). FDA staff formulate relevant questions for the committee and may also make presentations before the committee. This process is not limited to the immediate pre-licensure stage, however, as clinical data and related clinical issues may be referred to VRBPAC or other appropriate FDA advisory committees at any time during preclincal or clinical development.

Postlicensure Surveillance

The National Childhood Vaccine Injury Act of 1986 required that health-care providers report selected adverse events for certain childhood vaccines (i.e., DTP, polio, MMR) (24,25). Also, for all licensed vaccines, adverse reaction reports are entered into the Vaccine Adverse Events Reporting System (VAERS). These reports to VAERS can come from physicians, manufacturers, and others. CBER and Center for Disease Control and Prevention (CDC) staff monitor this data base for incidents of concern and patterns of events that appear to be vaccine-related in order to ensure that any potentially unsafe vaccine products or lots are expeditiously removed from the market and that other problems are identified and resolved.

This chapter reflects the author's assessment of the requirements for clinical trials for vaccines and is not intended to represent the official position of the FDA.

References

(1) Plotkin, S.L., and Plotkin, S.A., *A Short History of Vaccination*. Plotkin, S.A., and Mortimer, E.A., Jr. (eds.). Vaccines. W.B. Saunders, 1-7 (Philadelphia 1988).

(2) Peter, G., *Childhood Immunizations*, 327 New England Journal of Medicine 1794-1800 (1992).

(3) Mitchell, V.S., Philipose, N.M., and Sanford, J.P. (eds.), *The Children's Vaccine Initiative. Achieving the Vision,* Institute of Medicine. National Academy Press (Washington, D.C., 1993).

(4) Hopps, H.E., Meyer, B.C., and Parkman, P.D., *Regulation and Testing of Vaccines,* Plotkin, S.A., and Mortimer, E.A., Jr. (eds.). Vaccines. W.B. Saunders, 576-586 (Philadelphia 1988).

(5) Hinman, A.R., *Public Health Considerations*, Plotkin, S.A. and Mortimer, E.A., Jr. (eds.), Vaccines. W.B. Saunders, 587-611 (Philadelphia 1988).

(6) Combination Vaccines and Simultaneous Administration: Current Issues and Perspectives. Workshop sponsored by CBER/FDA, NVPO, NIAID/NIH, CDC, WHO. Bethesda, MD, July 28-30, 1993. (Proceedings to be published in the Annals of the New York Academy of Sciences).

(7) Edelman, R., *An Update on Vaccine Adjuvants in Clinical Trials*, 8 AIDS Res Hum Retroviruses 1409-1411 (1992).

(8) Dixon, D., Rida, W., Fast, P., and Hoth D., *HIV Vaccine Trials: Some Design Issues Including Sample Size Calculations*, 6 Journal of Acquired Immune Deficiency Syndromes 485-496 (1993).

(9) Esparza, J., Osmanov, S., Kallings, L., and Wigzell, H., *Planning for HIV Vaccine Trials: the World Health Organization Perspective*, 5 (suppl 2) AIDS S159-163 (1991).

(10) Blackwelder, W.C., Storsaeter, J., Olin P., and Hallander, H.O., *Acellular Pertussis Vaccines. Efficacy and Evaluation of Clinical Case Definitions*, 145 AJDC 1285-1289 (1991).

(11) Storsaeter, J., Hallander, H., Farrington, C.P., Olin, P., Mollby, R., and Miller, E., *Secondary Analyses of the Efficacy of Two Acellular Pertussis Vaccines Evaluated in a Swedish Phase III Trial*, 8 Vaccine 457-461 (1990).

(12) Ad Hoc Group for the Study of Pertussis Vaccines. Placebo-controlled Trial of Two Acellular Pertussis Vaccines in Sweden-Protective Efficacy and Adverse Events, 1 Lancet 955-960 (1988).

(13) Katz, S.P., *Effects of Malnutrition and Parasitic Infections on the Immune Response to Vaccines: Evaluation of the Risks Associated with Administering Vaccines to Malnourished Children*, Halsey, N.A., and de Quadros, C.A. (eds.). Recent Advances in Immunizaton. Pan American Health Organization, Scientific Publication #451, 81-89 (1983).

(14) Patriarca, P.A., Wright, P.F., and John, T.J., *Factors Affecting the Immunogenicity of Oral Poliovirus Vaccine in Developing Countries: Review*, 13 Reviews of Infectious Diseases 926-939 (1991).

(15) O'Brien, P.C., and Fleming, T.R., *A Multiple Testing Procedure for Clinical Trials*, 35 Biometrics 549-556 (1979).

(16) Geller, N.L. and Pocock, S.J., *Interim Analyses in Randomized Clinical Trials: Ramifications and Guidelines for Practitioners*, 43 Biometrics 213-223 (1987).

(17) Fleming, T.R., *Design Considerations for Clinical Trials,* Cancer Chemotherapy: Challenges for the Future. Vol. 3, K. Kimura, et. al. (eds.). Excerpta Medica (Tokyo 1988).

(18) Emerson, S.S. and Fleming, T.R., *Interim Analyses in Clinical Trials,* 4 Oncology 126-133 (1990).

(19) "Guideline for the Format and Content of the Clinical and Statistical Sections of an Application." July 1988. Prepared by the Center for Drug Evaluation and Research/FDA. Note: This 125-page guideline was prepared for drug approvals, but is still useful reading because many of the format, content, and statistical aspects of preparing a protocol, and collecting, analyzing, and presenting data are, in principle, very similar for drugs and biological products.

(20) Proceedings of "Practical Issues in Data Monitoring of Clinical Trials" meeting on 27-28 January 1992. 12 Statistics in Medicine 419-616 (1993).

(21) Black, S.B., Shinefield, H.R., Fireman, B., Hiatt, R., Polen, M., and Vittinghoff, E., *Efficacy in Infancy of Oligosaccharide Conjugate* Haemophilus influenzae *type b (HbOC) Vaccine in a United States Population of 61,080 Children,* 10 Pediatric Infectious Disease Journal 97-104 (1991).

(22) Santosham, M., Wolf, M., Reid, R. et al., *The Efficacy in Navajo Infants of a Conjugate Vaccine Consisting of* Haemophilus influenzae *Type b polysaccharide and* Neisseria meningitidis *Outer-membrane Protein Complex,* 324 New England Journal of Medicine 1767-72 (1991).

(23) Zollinger, W.D., *Meningococcol Meningitis.* Cruz, S.J. Jr. (Ed). Vaccines and Immunotherapy. Pergamon Press, Inc., 113-126 (New York 1991).

(24) CDC. Vaccine Adverse Event Reporting System - United States, 39 MMWR 730-733 (1990).

(25) National Childhood Vaccine Injury Act: Requirements for Permanent Vaccination Records and for Reporting of Selected Events After Vaccination, 37 MMWR 197-200 (1988).

Chapter 7:
Good Clinical Practices (GCP)

by Mark Mathieu
PAREXEL International Corporation

Because the FDA's approval of any biologic is based largely on clinical data, the agency has a vested interest in the conditions under which these data are obtained. Through a set of regulations and guidelines collectively known as "good clinical practices" (GCP), the FDA sets minimum standards for conducting clinical trials.

By identifying and defining the responsibilities of the key figures involved in clinical trials, the FDA's GCP regulations are designed to accomplish two primary goals: (1) to ensure the quality and integrity of the data obtained from clinical testing so that the FDA's decisions based on these data are informed and responsible; and (2) to protect the rights and, as much as possible, the safety of clinical subjects.

In reality, GCP is a term of convenience used by those in government and industry to identify a collection of related regulations that, when taken together, define the clinical trial-related responsibilities of the sponsor, the investigator, the monitor, and the institutional review board (IRB). These responsibilities are found primarily in four documents:

- a 1981 final regulation on the informed consent of clinical subjects;

- a 1981 final regulation on the responsibilities of IRBs;

- the 1987 IND Rewrite regulations, which define the responsibilities of the investigator and the sponsor (the IND Rewrite's provisions for sponsors and investigators supplants, at least temporarily, a 1977 proposed rule on the obligations of sponsors/monitors and a 1978 proposed rule on the obligations of investigators); and

- the 1988 *Guideline for the Monitoring of Clinical Investigations*, which outlines the monitor's responsibilities.

While these documents form the core of GCP, dozens of other FDA guidance documents provide more detailed information. Among these is a series of 28 *FDA Clinical Investigator Information Sheets* (see exhibit below), and a set of FDA compliance policy guidance manuals that specify how FDA inspectors ensure that clinical sponsors, monitors, and investigators are complying with GCP.

Responsibilities of the Sponsor

Federal regulations define "sponsor" as: "...a person who takes responsibility for and initiates a clinical investigation. The sponsor may be an individual or pharmaceutical company, governmental agency, academic institution, private organization, or other organization."

In general, the term "sponsor" refers to a commercial manufacturer that has developed a product in which it holds the principal financial interest. A sponsor may also be a physician, commonly called a "sponsor-investigator," which federal regulations define as "an individual who both initiates and conducts an investigation and under whose immediate direction the investigational drug is administered or dispensed."

The FDA defines sponsor responsibilities in Part 312, Subpart D in the *Code of Federal Regulations* (CFR), which states: "Sponsors are responsible for selecting qualified investigators, providing them with the information they need to conduct an investigation properly, ensuring proper monitoring of the investigation(s), ensuring that the investigation(s) is conducted in accordance with the general investigational plan and protocols contained in the IND, maintaining an effective IND with respect to the investigations, and ensuring that the FDA and all participating investigators are

FDA Clinical Investigator Information Sheets

- Acceptance of Foreign Data, IRB and Informed Consent Requirements
- Advertising for Study Subjects
- Clinical Investigator Regulatory Sanctions
- Clinical Investigators Unaffiliated with an Institution with an IRB
- Continuing Review
- Cooperative Research
- Emergency Use of an Investigational Drug
- FDA Inspections of Clinical Investigators
- FDA Institutional Review Board Inspections
- Guidance for the Emergency Use of Unapproved Medical Devices
- Guidance on Significant and Nonsignificant Risk Device Studies
- Informed Consent and the Clinical Investigation
- Investigational Drug Use in Patients Entering a Second Institution
- Investigational Use of Marketed Products
- IRBs and Medical Devices
- Non-Local IRB Review
- Payment to Research Subjects
- Placebo-Controlled and Active Controlled Drug Study Designs
- Required Recordkeeping in Clinical Investigations
- Significant Difference in HHS and FDA Regulations for IRBs and Informed Consent
- Sponsor-Clinical Investigator - IRB Interrelationship
- Treatment Use of Investigational Drugs
- Waiver of IRB Requirements

Source: FDA

promptly informed of significant new adverse effects or risks with respect to the drug." Sponsor responsibilities can be divided into the following general areas:

• selecting investigators and monitors;

• informing investigators;

• reviewing ongoing investigations;

• recordkeeping and record retention; and

• ensuring disposition of unused biologic supplies.

Selecting Investigators and Monitors   *Investigator Selection.* The sponsor must select investigators—physicians and other professionals contracted by the sponsor to conduct the clinical study, including supervising the administration of the biologic to human subjects—qualified by training and experience as appropriate experts to investigate the biologic. Sponsors may ship investigational product to these investigators only.

To ensure that an investigator is qualified, the sponsor must obtain certain information from the investigator:

• A Completed and Signed Statement of Investigator Form (Form FDA-1572). This form contains information about the investigator, the site of the investigation, and the subinvestigators—research fellows and residents—who assist the investigator in the conduct of the investigation. By signing the form, the investigator also pledges: (1) to conduct the study in accordance with the clinical protocol(s) and to take proper actions should deviations be needed; (2) to comply with all requirements regarding the obligations of clinical investigators (as described later in this chapter) and other relevant requirements; (3) to personally conduct or supervise the described investigation; (4) to inform patients, or any persons used as controls, that the biologic is being used for investigational purposes and to ensure that the requirements relating to obtaining informed consent and IRB review and approval (as described elsewhere in this chapter) are met; (5) to report to the sponsor adverse experiences that occur in the course of the investigation in accordance with regulatory requirements; (6) to read and understand the information in the investigator's brochure, including the potential risks and side effects of the biologic; and (7) to ensure that all associates, colleagues, and employees assisting in the conduct of the studies are informed about their obligations in meeting the above commitments. Through the form, the investigator also pledges that an IRB operating in compliance with regulatory requirements (described later in this chapter) will be responsible for the initial and continuing review and approval of the clinical investigation. In addition, the investigator promises to report to the IRB all changes in the research activity and all unanticipated problems involving risks to human subjects and to not make any such changes without IRB approval, except when necessary to eliminate apparent immediate hazards to the human subjects.

- Curriculum Vitae. The sponsor must obtain a curriculum vitae or other statement of qualifications of the investigator showing the education, training, and experience that qualify the investigator as an expert in the clinical investigation of the biologic.

- Clinical Protocol. For Phase 1 investigations, the sponsor must obtain from the investigator a general outline of the planned investigation, including the estimated duration of the study and the maximum number of subjects that will be involved. For Phase 2 or 3 investigations, the sponsor must obtain an outline of the study protocol, including an approximation of the number and characteristics of investigational subjects and controls, the clinical uses to be investigated, the kinds of clinical observations and laboratory tests to be conducted, the estimated duration of the study, and copies or a description of case report forms to be used.

Selecting Monitors and Monitoring the Clinical Trial. Sponsors are required to monitor clinical investigations to ensure: (1) the quality and integrity of the clinical data derived from clinical trials; and (2) that the rights and safety of human subjects involved in a clinical study are preserved. The monitoring function may be performed by the sponsor or its employees, or may be delegated to a contract research organization (CRO).

Specific FDA recommendations on proper monitoring duties and procedures are provided in the agency's *Guideline for the Monitoring of Clinical Investigations* (January 1988). In this document, the FDA identifies six different monitoring responsibilities:

- Selection of a Monitor.   According to the guideline, a sponsor may designate one or more appropriately trained and qualified individuals to monitor the progress of a clinical investigation. Physicians, clinical research associates, paramedical personnel, nurses, and engineers may be acceptable monitors depending on the type of product involved in the study.

- Written Monitoring Procedures.   A sponsor should establish written procedures for monitoring clinical investigations to assure the quality of the study, and to assure that each person involved in the monitoring process carries out his or her duties.

- Preinvestigation Visits.   Through personal contact between the monitor and each investigator, a sponsor must assure that the investigator, among other things, clearly understands and accepts the obligations involved in undertaking a clinical study. The sponsor also must determine whether the investigator's facilities are adequate for conducting the investigation and whether the investigator has sufficient time to honor his or her responsibilities in the trial.

- Periodic Visits.   A sponsor must assure, throughout the clinical investigation, that the investigator's obligations are fulfilled and that the facilities used in the clinical investigation are acceptable. The monitor must visit the clinical site frequently enough to provide such assurances.

- Review of Subject Records.   A sponsor must assure that safety and effectiveness data submitted to the FDA are accurate and complete.

The FDA recommends that the monitor review individual subject records and other supporting documentation and compare those records with the reports prepared by the investigator for submission to the sponsor.

- Record of On-Site Visits. The monitor or sponsor should maintain a record of the findings, conclusions, and actions taken to correct deficiencies for each on-site visit.

Informing Investigators   The sponsor is responsible for keeping all investigators involved in the clinical testing of its biologic fully informed about the investigational product and research findings. Before the investigation begins, a sponsor must provide participating clinical investigators with an investigator's brochure (see Chapter 4), which provides a description of the product and summaries of its known pharmacological, pharmacokinetic, and biological characteristics; potential adverse effects as indicated by animal tests; and, if available, data on clinical use.

Once clinical trials begin, regulations require that sponsors "keep each participating investigator informed of new observations discovered by or reported to the sponsor on the [biologic], particularly with respect to adverse effects and safe use." This information may be distributed through periodically revised investigator brochures, reprints or published studies, reports or letters to clinical investigators, or other appropriate means. Important safety information must be relayed to investigators and the FDA through written or verbal IND safety reports (see Chapter 4).

Review of Ongoing Investigations   There are many reasons why the FDA requires sponsors to closely monitor the conduct and progress of their clinical trials. Investigator noncompliance and unreasonable and significant drug risks are two of the most important reasons.

If a sponsor discovers that an investigator is not complying with his or her commitments in Form FDA-1572, the general investigational plan, or other relevant regulatory requirements, the firm must either secure compliance or discontinue product shipments to the investigator and terminate the investigator's participation in the investigation. If the latter course is chosen or is necessary, the sponsor must require that the investigator return or dispose of the product in accordance with applicable requirements and must report this action to the FDA.

The sponsor must review and evaluate safety and effectiveness data as it is supplied by the investigator. In addition to providing important safety information through IND safety reports, the sponsor must supply to the FDA annual reports on the progress of the investigation.

Sponsors finding that their products present unreasonable and significant risks to subjects must: (1) discontinue those investigations that present the risks; (2) notify the FDA, all IRBs, and all investigators who have at any time participated in the investigation that the study is being discontinued; (3) assure the disposition of all outstanding stocks of the biologic; and (4) furnish the FDA with a full report of its actions.

Recordkeeping and Record Retention A sponsor must maintain adequate records showing the receipt, shipment, or other disposition of the investigational product. The records must include, as appropriate, the name of the investigator to whom the biologic is shipped and the date, quantity, and batch or code mark of each such shipment. Regulations require a sponsor to retain the records and reports for two years after either its marketing application is approved or its notification to the FDA that product shipment and delivery have been discontinued.

Disposition of Unused Biologic Supplies The sponsor must ensure the return of all unused supplies of the biologic from each investigator whose participation is discontinued or eliminated. The sponsor may authorize alternative plans, provided these do not expose humans to risks from the product.

Responsibilities of Investigators
A clinical investigator is the individual who actually conducts, or who is the responsible leader of a team of individuals that conducts, a clinical investigation. It is under the immediate direction of this individual that the product is administered or dispensed to a clinical subject.

Federal regulations state that an "... investigator is responsible for ensuring that an investigation is conducted according to the signed investigator statement, the investigational plan, and applicable regulations; for protecting the rights, safety, and welfare of subjects under the investigator's care; and for the control of [biologics] under investigation." As part of the investigator's role in protecting the rights of clinical subjects, he or she must obtain the informed consent of all human subjects to whom the product is administered. Specific investigator responsibilities detailed in GCP provisions include:

Control of the Product. The investigator can administer the product only to subjects under his or her personal supervision or under the supervision of a subinvestigator. Regulations do not allow the investigator to supply the biologic to persons not authorized to receive it.

Recordkeeping and Record Retention. The investigator must keep adequate records regarding the disposition of the product and subject case histories recording all observations and data pertinent to the investigation. These records must be kept for two years after either a marketing application's approval or a sponsor has discontinued an IND and so notified the FDA. The FDA must be allowed access to these records.

Investigator Reports. The investigator must provide to the sponsor: (1) annual reports on the progress of the clinical investigations; (2) safety reports on all adverse effects that may reasonably be regarded as caused by, or probably caused by, the biologic; and (3) a final report shortly after the completion of the investigator's participation—FDA officials indicate that completed case report forms on all subjects will suffice.

Assurance of IRB Review. The investigator must assure that an IRB complying with regulatory requirements will be responsible for the initial and continuing review and approval of the proposed clinical study. He or she must also promptly report to the IRB all changes in the research activity and all unanticipated problems involving risks to human subjects. The investigator must not make any changes in the research without IRB approval, except when necessary to eliminate apparent and immediate hazards to human subjects.

Handling of Controlled Substances. If the investigational product is subject to the Controlled Substances Act, the investigator must take adequate precautions to prevent theft or diversion of the substance.

The Institutional Review Board (IRB)

The IRB's function is to see that risks to clinical subjects are minimized and that the subjects are adequately informed about the clinical trial and its implications for their treatment. In doing so, the IRB's authority goes beyond just reviewing proposed clinical protocols and the ongoing trial. Although the board's main concern is not the adequacy of study design, the board can order that a trial be modified for safety or other reasons.

The IRB itself must consist of at least five persons, each of whom is chosen by the institution. Board members must be judged to have the professional competence necessary to review specific research activities and to have the ability to assess the acceptability of proposed research in terms of institutional commitments and regulations, applicable law, and standards and practice.

IRB members often are physicians, pharmacologists, and administrative managers from the parent institution. At least one board member, however, must have a primary interest in a nonscientific area such as law, ethics, or religion. Federal regulations also include several other requirements that are designed to ensure the independence of the board and guard against conflicts of interest.

Generally, biologic sponsors have little, if any, direct contact with an IRB. In fact, the FDA openly discourages drug sponsors from communicating directly with IRBs. The investigator heading the study at a particular institution will usually act as a liaison, and will present the study plans for IRB consideration and approval. Through past experience, the investigator will usually be familiar with the particular concerns and priorities of an IRB and is, therefore, better prepared to deal with its members.

Aside from safety concerns, an IRB may address several issues—including specific standards of the institution, state, and locality—in evaluating a certain study. Any research program the board does approve, however, must meet several criteria specified in FDA regulations:

• risks to subjects must be minimized;

• risks to subjects must be reasonable in relation to the anticipated benefits and the importance of the knowledge that may be expected to be gained;

• subject selection must be equitable;

• informed consent must be sought from each prospective subject or the subject's legally authorized representative;

• informed consent must be appropriately documented;

• when appropriate, the research plan must make adequate provisions for monitoring the data collected to ensure the safety of subjects; and

• when appropriate, there must be adequate provisions to protect the privacy of subjects and to preserve the confidentiality of data.

As are sponsors, monitors, and investigators, IRBs are subject to reporting and recordkeeping requirements. The board must retain minutes of meetings, copies of all research proposals reviewed, sample consent documents, correspondence with investigators, board procedures, and other documents. IRB meetings and records are subject to FDA inspections, and an institution may be disqualified from conducting clinical studies if FDA inspectors find that its IRB has violated GCP requirements.

Informed Consent

Informed consent is a concept designed to ensure that patients do not enter a clinical trial either against their will or without an adequate understanding of their medical situation and the implications of the clinical study itself. Federal regulations dictate that, except under special circumstances, "...no investigator may involve a human being as a subject in research ... unless the investigator has obtained the legally effective informed consent of the subject or the subject's legally authorized representative. An investigator shall seek such consent only under circumstances that provide the prospective subject or the representative sufficient opportunity to consider whether or not to participate and that minimizes the possibility of coercion or undue influence. The information that is given to the subject or the representative shall be in language understandable to the subject or the representative."

Clearly, informed consent implies an informed patient. Any subject volunteering for the study must be fully aware of his/her medical condition, alternative treatments, and the purpose of and risks involved in the clinical study. Federal regulations regarding the protection of clinical subjects state that, at the minimum, the following "basic elements of informed consent" must be provided to clinical subjects before involving them in the trial:

- a statement that the study involves research, an explanation of the purposes of the research and the expected duration of the subject's participation, a description of the procedures to be followed, and identification of any procedures that are experimental;

- a description of any reasonably foreseeable risks or discomforts to the subject;

- a description of any benefits that the subject or others may reasonably expect from the research;

- a disclosure of appropriate alternative procedures or courses of treatment, if any, that might be advantageous to the subject;

- a statement that describes the extent, if any, to which confidentiality of records identifying the subject will be maintained and that notes the possibility that the FDA may inspect the records;

- an explanation as to whether any compensation or medical treatments are available if injury occurs during research involving more than minimal risk, and, if so, what the treatments and/or compensation consist of, or where further information may be obtained;

- the identity of the person to contact for answers to pertinent questions about the research and research subject's rights, and the person to contact if the subject suffers a research-related injury; and

- a statement that participation is voluntary, that refusal to participate will involve no penalty or loss of benefits to which the subject is otherwise entitled, and that the subject may discontinue participation at any time without penalty or loss of benefits to which the subject is otherwise entitled.

When appropriate, one or more of the following must also be provided to subjects:

• a statement that a particular treatment or procedure may involve risks to the subject (or to the embryo or fetus, if the subject is or may become pregnant) that are currently unforeseeable;

• anticipated circumstances under which the subject's participation may be terminated by the investigator without regard to the subject's consent;

• any additional costs to the subject that may result from participation in the research;

• the consequences of a subject's decision to withdraw from the research and procedures for ordering termination of participation by the subject;

• a statement that significant new research findings that may affect the subject's willingness to continue his or her participation will be provided to the subject; and

• the approximate number of subjects involved in the study.

In most cases, informed consent must be obtained by having the subject or the subject's representative sign a written consent form that has been approved by the IRB. Unless the IRB waives the informed consent requirements due to absence of risk, the consent form may take one of two forms: (1) a written consent document that embodies the basic elements of informed consent and that may be read to the subject or the subject's representative, who is then given adequate opportunity to read it before signing; or (2) a "short form" written consent document stating that the basic elements of informed consent have been presented orally to the subjects or the subject's representative. If the short form is used, there are several other requirements: there must be a witness to the oral presentation; the IRB shall approve a written summary of what will be said to the subject or the representative; the witness must sign both the short form and a copy of the summary; the person obtaining the consent must sign a copy of the summary; and the subject must be given both the summary and a copy of the consent form.

While the investigator is directly responsible for obtaining a subject's informed consent and seeing that the subject is truly informed, the IRB and the sponsor/monitor also play roles in ensuring that informed consent requirements are met.

Chapter 8

The Product License Application (PLA)

by Al Ghignone
President
AAG & Associates

Before a company can market a new biological product legally in the United States, the sponsor must hold two government licenses: a product license and an establishment license. To obtain these licenses, the sponsor files both a product license application (PLA) and an establishment license application (ELA) (see Chapter 9 for a detailed analysis of ELAs).

In the PLA, a biological product sponsor submits thousands of pages of nonclinical and clinical data, chemical and biological information, and product manufacturing descriptions. The submission must allow CBER reviewers to make three principal determinations:

1. whether the biologic is safe and effective in its indicated use and whether the benefits of using the product outweigh the risks.

2. whether the biologic's proposed labeling is appropriate.

3. whether the methods used in manufacturing and quality control are adequate to preserve the biologic's identity, strength, quality, potency, and purity.

A Short History of the Licensing Process for Biologics

Given that government regulation of biologics has historically focused on the product manufacturing process, it is not surprising that the ELA has a considerably longer history than the PLA. Congress was first spurred to regulate the biological industry in 1901, when ten children died after being treated with diphtheria antitoxin that had been contaminated with tetanus.

But in establishing regulatory controls to prevent the contamination of the biological products of that day—essentially vaccines and antitoxins—Congress had to work within the limitations of existing scientific knowledge and technology. At that time, it was difficult, in many cases impossible, to identify component parts of any biological product or to detect the presence of pathogens and other contaminants. The absence of sensitive assays for identifying, and purification processes for separating, biological contaminants left researchers to use crude immunological and *in vivo* tests.

This situation was complicated by the fact that the production processes for traditional biologics—usually involving human or animal extracts—were highly susceptible to contamination. The reality that contaminants were often infectious materials or toxins amplified the threat.

At that time, regulating production facilities seemed to be the only mechanism likely to control the quality of biological products. Consequently, Congress passed the Biological Control Act of 1902, which required that biologics in interstate commerce be manufactured in facilities holding a valid establishment license. Interestingly, the statute mandated no government review or sanction of the products themselves, only that the establishments manufacturing and preparing the products meet specific criteria and permit the inspection of their facilities.

Not until 1944, when Congress modified the statute, did federal law require the licensure of products as well. Under the revision, both establishments and products must "meet standards designed to insure the continued safety, purity, and potency of such products, prescribed in regulation."

Formatting Requirements for PLAs

The FDA has established no specific formatting requirements for PLAs. This is in sharp contrast to new drug applications, for which the agency has detailed formatting standards. Although the lack of a uniform PLA format has

146

caused industry concern in the past, significant differences between various license applications—and biological products themselves—have slowed efforts to standardize PLA formats.

CBER does make available a series of PLA application forms that identify basic PLA submission requirements for different types of products (see below). However, most of these PLA forms apply to blood products, vaccines, and other, more traditional biological products. The center has not yet developed forms for many of the more advanced biological products, including several therapeutics.

In most cases, PLA forms identify submission requirements by posing a series of questions that the sponsor must answer in the application. Essentially, the lack of a standard format is a function of the diversity of questions posed by different PLA forms. Often, applicants base the formats of their applications on the sequence of questions specified in the PLA form.

Given the absence of a standard PLA format, this chapter outlines a general format that may be used for product license applications. In addition to this suggested format, sponsors are advised to consult other information sources that will assist in the preparation of the PLA:

CBER Guidelines and Points-To-Consider Documents. CBER has developed several guidelines and points-to-consider documents, including *Points to Consider in Human Somatic Cell Therapy and Gene Therapy* (August 1991) and *Points to Consider in the Design and Implementation of Field Trials for Blood Grouping Reagents and Anti-Human Globulin* (March 1992). For a complete listing of these guidelines and points-to-consider documents, see Appendix 1.

Product License Application Forms. As listed below.

Federal Regulations. Some guidance is available in FDA regulations, specifically Title 21: Code of Federal Regulations (CFR): Parts 201-202 - Labeling and Advertising; Part 211 - Good Manufacturing Practices (GMP); Part 312 - IND; Parts 600-610 - General Licensing Provisions; and Parts 620-680 - Product Standards.

In general terms, the PLA consists of reports of all investigations sponsored by the applicant, and all other information pertinent to an evaluation of the product's safety, effectiveness, potency, and purity. The suggested format for presenting this information in the PLA consists of 16 distinct sections:

Section 1 - Application Form
Section 2 - Cover Letter
Section 3 - Table of Contents
Section 4 - Introduction
Section 5 - Summary
Section 6 - Manufacturing and Controls Section
Section 7 - Preclinical Section
Section 8 - Clinical Section
Section 9 - Stability Section
Section 10 - Facilities: Systems and Design
Section 11 - Labeling Section
Section 12 - Product Samples
Section 13 - Responsible Personnel
Section 14 - Environmental Assessment
Section 15 - Certification of Regulatory Compliance
Section 16 - Other

Although these sections do not correspond item for item with each PLA application form, they do comprise the totality of information requested in a PLA. For example, the table of contents and introduction sections in this format comprise what many PLA application forms request in the "General Information" section.

Before addressing each section individually, it is worth emphasizing the importance of the first five sections. Applicants frequently overlook their significance, perhaps because they are not technical sections or because they often are not seen as critical.

For several reasons, this is unfortunate. First, these sections are among the few in the entire application that each member of the PLA licensing committee receives for review. Additionally, because these sections—particularly the cover letter, summary, and introduction—provide information in abridged form, they are also likely to be read thoroughly by each committee member.

The sections are also important because they represent, in some ways, the applicant's "opening argument" for its product. In the summary section, for instance, the sponsor is granted what some view as a greater editorial license not available in any of the PLA's other sections. Such views aside, the sponsor must use these sections to frame and build its case for the new biologic's safety and effectiveness.

Section 1 - Application Form CBER makes available PLA application forms that identify submission requirements for specific product types. In some cases, such as applications for blood and vaccine products, these forms must be signed by the sponsor's responsible head and must accompany the PLA submission.

In situations in which CBER has developed only draft application forms, the forms are recommended, but are not required in PLA submissions. Such is the case for several PLA application forms for therapeutic products.

As of mid-1993, CBER had approximately 20 application forms, including the following:

- Application for Allergenic Extracts (PLA Form 3213);
- Application for Bacterial Vaccines and Antigens (PLA Form 3212);
- Application for Viral and Rickettsial Products (PLA Form 3211);
- Application for License for the Manufacture of a Human Plasma Derivative (PLA Form 3214);
- Application for Human Immunodeficiency Virus for In-Vitro Diagnostic Use (PLA Form 3314);
- Application for Source Plasma (PLA Form 2600);
- Application for Blood Grouping Reagents (PLA Form 3066);
- Application for Reagent Red Blood Cells (PLA Form 3086);
- Application for Whole Blood and Blood Components (PLA Form 3098);
- Application for Red Blood Cells (PLA Form 3098A);
- Application for Plasma (PLA Form 3098B);
- Application for Platelets (PLA Form 3098C);
- Application for Cryoprecipitated Antihemophilic Factor (PLA Form 3098D);
- Application for Cytopheresis Products (PLA Form 3098E);

- Draft Application for License for the Manufacture of Interferon (draft);
- Draft Application for License of Recombinant DNA-Derived Biological Products (draft);
- Application for License for the Manufacture of Monoclonal Antibody Therapeutic Products (draft);
- Application for the Manufacture of Anti-Human Globulin (PLA Form 3096); and
- Application for Therapeutic Exchange Plasma (PLA Form 2600b)

Section 2 - Cover Letter   In the cover letter, sponsors often supply the FDA with much of the basic information requested in the PLA application form (e.g., sponsor name and address, etc.) and additional information. When no application form is available for the particular product, the cover letter serves as that form. In either case, the cover letter should provide at least seven types of information:

Name and Address of the Applicant and Others. The cover letter should provide the name and address of the sponsor. If the sponsor is using outside contractors or manufacturing sites at other locations, the cover letter should provide their addresses and identify their functions.

Responsible Head. The cover letter should also identify the sponsor's responsible head, who exercises control in all matters relating to an establishment's regulatory compliance. In doing so, the responsible head acts with the full authority to represent the manufacturer in all matters relative to CBER. As the sponsor's main contact with CBER, the individual should sign all regulatory correspondence.

Product Name. The sponsor should provide the trade and generic name of the product in the cover letter.

Reason for Submission. The cover letter should identify the type of application being submitted (e.g., original submission, supplement, amendment, etc.).

Information Contained in the Submission. In the cover letter, the sponsor should identify what information is contained in the submission. For example:

Volume 1   - Introduction and Summary
Volume 2   - Manufacturing and Control Section
Volume 3   - Preclinical Section
Volumes 4-7   - Clinical Section

Agreements With the FDA. If the sponsor has reached any agreements with CBER relevant to the PLA, this information should be included in the cover letter. Given the quantity of applications under review within CBER and the fact that such agreements often are made months in advance, reviewers might not recall the existence or details of such agreements. Reviewer turnover is another factor that makes recounting these agreements good working practice.

Other Documents Relating to the Submission. To alert CBER reviewers to other documentation that must be referenced during the PLA review, the sponsor should note in the cover letter other documents associated with the product, such as INDs, PLAs, ELAs, and master files.

Section 3 - Table of Contents Perhaps the single most important factor in a PLA's "user friendliness" is the speed and ease with which a reviewer can find information during the review process. Since it can influence the speed and efficiency of the review as well, the manner in which the applicant indexes PLA information is of central importance. Applicants can use the format below for indexing the PLA:

Item	Subject	Volume	Page
1	PLA Application Form	1	—
2	Cover Letter	1	—
3	Table of Contents	1	1.0001
4	Introduction	1	1.0005
5	Summary	1	1.0011

In the page column, the number to the left of the decimal point represents the application's volume number, while the number to the right refers to the page within that volume containing the relevant section. In practice, the table of contents is typically far more detailed, with each section broken down into specific subparts.

Section 4 - Introduction   The introduction is just that: an introduction to the sponsor, the product and its indicated use, and the application. This is one of the sections that all members of the PLA review committee will read to acquaint themselves with the sponsor and the product. The introduction should include the following information:

- the sponsor's name and address;
- the manufacturing establishment's name, address, and if applicable, license number;
- the name, address, and, if applicable, license numbers of other contractors;
- the product's proper name and all other names used for the product; and
- the product's proposed indication.

Section 5 - Summary   In many ways, the PLA summary is a condensed version of the entire application. The summary serves as a guide to the full PLA, explaining the application's intent—to prove the biologic's safety and effectiveness for a particular indication—and highlights the studies and evidence supporting the biologic's safety and effectiveness.

The summary's importance cannot be overstated. In this section, the applicant can state and argue its case for the product's approval. A well-prepared summary, which should include a straightforward description of the product and its manufacturing technology and adverse and beneficial effects, can build CBER's confidence in the applicant, the validity of the PLA's information, and the product itself.

As mentioned previously, the summary is one of the few sections that all members of the PLA licensing committee review. Therefore, this document can be pivotal in establishing a foundation for product approval. It is also the only PLA section that provides the sponsor with somewhat greater latitude in describing the benefits of its product.

The most detailed discussion of an application summary is found in a 1985 FDA guideline entitled *Draft Guideline for the Format and Content of an Application Summary*. Although this guideline applies to NDAs, its general principles are equally applicable to PLA summaries. According to this document, an application should "contain a summary, ordinarily 50 to 200 pages in length, that integrates all of the information in the application and provides reviewers in each review area, and other agency officials, with a good general understanding of the drug product and of the application. The summary should discuss all aspects of the application and should be written in approximately the same level of detail required for publication in, and meet the editorial standards generally applied by, refereed scientific and medical journals .... To the extent possible, data in the summary should be presented in tabular and graphic forms .... The summary should comprehensively present the most important information about the ... product and the conclusions to be drawn from this information. The summary should avoid any editorial promotion of the ... product, i.e., it should be a factual summary of safety and effectiveness data and a neutral analysis of these data. The summary should include an annotated copy of the proposed labeling, a discussion of the product's benefits and risks, a description of the foreign marketing history of the drug (if any), and a summary of each technical section."

Section 6 - Manufacturing and Controls Section   Historically, the government's regulation of biologics has focused on the manner in which biological products are manufactured. It is not surprising, then, that the vast majority of information requested in CBER's various PLA application forms pertains to product manufacturing issues. Of the 14 pages of CBER's *Draft Application for License of Recombinant DNA Derived Biological Products*, for instance, virtually nine full pages pose manufacturing-related questions focusing on issues such as how the product is derived, specifications for the purified bulk product, the final product composition, and the manufacture of the final product. Meanwhile, just over one page is used to pose questions regarding the nonclinical and clinical sections, labeling, samples and protocols, records, environmental impact, and establishment data.

While both the PLA and ELA provide manufacturing information, each license application plays a different role. The PLA is designed to provide CBER with a complete description of the manufacture and control of the biological product. Conversely, the ELA provides a complete description of the facility in which the biological product is manufactured (see Chapter 9). Together, the applications provide CBER with a complete picture of the production facility and the product manufacturing process.

Although applicants are not required to do so, some companies provide a brief description of the production facility in the PLA's manufacturing and controls section. This provides the PLA licensing committee with a more complete account (facility and process) of the biological product's manufacture.

The majority of questions or problems that arise during PLA reviews relate to this portion of the license application. In this section, the sponsor fully describes the composition, manufacture, and specifications of two principal entities:

1. the biological substance, also referred to as the active moiety or purified bulk product.
2. the biological product, which is the final product.

Biological Substance. The PLA must provide a complete description of the biological substance, including:

- the biological substance's chemical, physical and biological characteristics and stability;
- the name and address of the substance's manufacturer;
- the method of manufacture and purification;
- the cell line and cell-banking procedures, and the methods for maintaining and validating the purity, identity, and potency of the cell line;
- the process controls used in manufacturing and packaging; and
- the specifications and analytical methods necessary to assure the identity, quality, potency, and purity of the biological substance (the term analytical methods is used generically and includes chemical, biological and other testing methods).

Biological Product. For the final product, the PLA must provide:

- a list of all components used in the manufacture of the biological product;
- a statement of the composition of the biological product;
- a statement of the specifications and analytical methods for each component;
- the manufacturer's name and address;

- a description of the manufacturing and packaging procedures and in-process controls used for the biological product; and
- the specifications and analytical methods necessary to assure the identity, potency, quality, and purity of the biological product.

The goals of a PLA's manufacturing section are to demonstrate to the FDA that the manufacturing operation can:

1. obtain the desired protein.
2. remove contaminants (host or inducer).
3. remove contaminants caused by the process.
4. avoid or remove product variants.
5. preserve the integrity of the protein.
6. maintain *in vivo* consistency.

The information presented to the agency should flow logically and be easy to read, a challenge considering the complexity of the manufacturing process. However, applicants can aid the review process by using flow charts accompanied by narrative explanations.

The following discussion provides a more detailed analysis of the information required in this section of the PLA. The format suggested can be adapted for use in license applications for recombinant and natural products.

For purposes of analysis, the manufacturing process is perhaps best divided into six steps, each of which should comprise a distinct element of the manufacturing section:

1. development and characterization of the production system.
2. preparation of the purified bulk product.
3. preparation of the formulated bulk product.
4. preparation of the final product.
5. in-process and final-product testing.
6. lot numbering.

Each of these six sections should have an introduction followed by a flow chart of the corresponding process. The flow charts should be accompanied by in-depth narratives of the relevant steps in the process, along with the specifications and test methods.

Biologics Development: A Regulatory Overview

Development and Characterization of the Production System. The production system may be divided into three (3) main components:

- molecular biology of the gene;
- characterization of the cell line; and
- manufacturer's working cell bank.

The gene and cell lines comprise the major production system components. The manufacturer's working cell bank is the means for perpetuating the components of the production system.

Molecular Biology of the Gene. The gene and its host (e.g., yeast, *e. coli*) comprise the expression system for the protein product. This section should consist of a general introduction and flow chart, and should provide information in three areas:

1. Segment Coding (gene sequence). The PLA should describe how the segment coding for the desired protein is prepared. The description should include information on the cell type and the origin of the source material.
2. Vector. The applicant should describe the method used for constructing the vector—the virus that delivers the gene sequence to the host system. The discussion should identify the source and function of the vector's component parts.
3. Host Cell System. The PLA should describe the host cell system's source, phenotype, and genotype. This section should also discuss the mechanism of transfer, the copy number, and the physical state of the vector inside the host cell.

The applicant must also demonstrate the characterization of the segment coding, vector, and host cell systems. This characterization allows the sponsor to establish specifications and test procedures to control these systems on an ongoing basis. The characterization, specifications, and test procedures should be included in this section.

Characterization of the Cell Line. In the PLA, the sponsor must identify the source, and provide full characterization (i.e., specifications, etc.), of the cell line. This section should provide descriptions of the following:

- the history and genealogy of the cell line;
- the master cell bank and the manufacturer's working cell bank (see discussion below);
- the storage of the master cell bank and the manufacturer's working cell bank;
- the culture medium; and
- growth characteristics *in vitro.*

The applicant should also include in this section information on testing conducted for adventitious agents.

If monoclonal antibodies are the final product or are used in-process (i.e, in the purification process), the sponsor must conduct a complete cell line characterization. The following information is required for such a product:

- the source, name, and characterization of the parent cell line;
- the species, animal strain, characterization, and tissue of origin;
- the identification and characterization of the immunogen;
- a description of the immunization scheme;
- a description of the screening procedure;
- a description of cell cloning procedures; and
- a description of the seed lot used for establishing and maintaining the primary master cell bank and manufacturer's working cell bank.

Manufacturer's Working Cell Bank. In most cases, the cell line used to produce a biologic will derive from a larger group of cells called the master cell bank, or MCB. In the PLA, the MCB must be fully characterized, particularly with respect to identity, stability, and microbial contaminants. Both specifications and test methods must be established during the characterization process.

The manufacturer's working cell bank (MWCB) is an aliquot—sample—of the MCB, and is the cell line used to produce the biological product. In the

PLA, the MWCB must be well characterized to demonstrate that, during the expansion of the MCB, genetic stability was maintained.

Additional information regarding the testing of cell lines is available from CBER's *Points to Consider Document on the Characterization of Cell Lines Used to Produce Biologicals* (revised May 1993). This document

Manufacturing and Controls

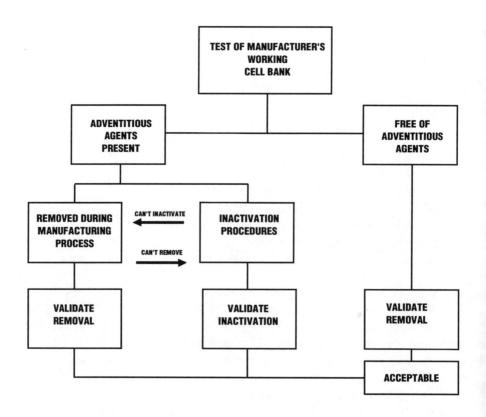

encourages sponsors of certain products to refer to other center publications as well: "Additional information concerning the testing of cell lines used to produce monoclonal antibodies and recombinant DNA technology products for *in vivo* and select *in vitro* use may be found in the 'Points to Consider in the Manufacture and Testing of Monoclonal Antibody Products for Human Use (June 1987)' (now under revision), 'Points to Consider in the Production and Testing of New Drugs and Biologicals Produced by Recombinant DNA Technology (April 1985)' and the supplement to the recombinant DNA Points to Consider, 'Nucleic Acid Characterization and Genetic Stability (1992).' If a cell line or cells are to be returned into humans to produce its biological product(s) *in vivo*, then the 'Points to Consider in Human Somatic Cell Therapy and Gene Therapy (1991)' and also the 'Points to Consider in the Collection, Processing and Testing of Ex-Vivo-Activated Mononuclear Leukocytes for Administration to Humans (1989)' should be consulted."

Preparation of the Purified Bulk Product.   The purified bulk product is the active component of the biologic. Also referred to as the biological substance, drug substance, or active ingredient, the purified bulk product accounts for a product's pharmacological activity.

The manufacture of the purified bulk product includes the following processes:

• fermentation, cell culture;
• harvesting; and
• column chromatography/purification.

This element of the PLA's manufacturing and controls section should include information regarding the preparation of the purified bulk product, which may be formatted into the following sections:

A. Introduction.
B. Flow chart.
C. Process description.
D. Raw materials and equipment.
E. In-process controls and specifications.

    F.   Validation summary.

    G. Packaging and storage.

    H. Stability.

Preparation of the Formulated Bulk Product. The formulated bulk product is the final biologic in bulk form, awaiting placement into the final container. When providing information on the preparation of the formulated bulk product, applicants can use a format similar to that of the previous section:

    A. Introduction.

    B. Flow chart.

    C. Process description.

    D. Equipment, raw materials, and solutions.

    E. In-process controls and specifications.

    F.   Validation summary.

    G. Packaging and storage.

    H. Stability.

Preparation of the Final Product. The final product is the biologic in the final packaged system marketed by a company. To provide continuity with the preceding sections, information on the preparation of the final product should be presented in a similar format:

    A. Introduction.

    B. Flow chart.

    C. Process description (batch records).

    D. Quantitative/qualitative compositions.

    E. Preservatives/diluents.

    F.   Equipment and raw materials.

    G. Preparation/processing of containers/closures.

    H. In-process controls and specifications.

    I.   Reprocessing/rework procedures.

    J.   Validation summary.

    K. Storage of final product.

    L.   Stability.

Production of Final Product

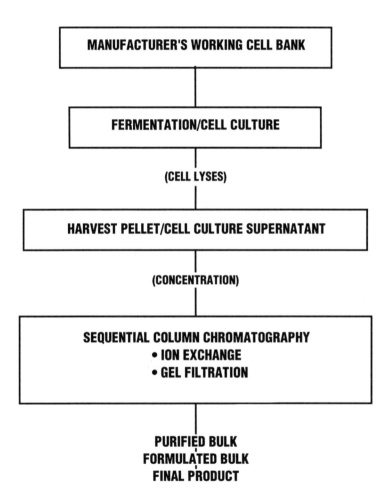

MANUFACTURER'S WORKING CELL BANK

FERMENTATION/CELL CULTURE

(CELL LYSES)

HARVEST PELLET/CELL CULTURE SUPERNATANT

(CONCENTRATION)

SEQUENTIAL COLUMN CHROMATOGRAPHY
• ION EXCHANGE
• GEL FILTRATION

PURIFIED BULK
FORMULATED BULK
FINAL PRODUCT

In-Process and Final-Product Testing. In this section, the applicant describes the testing conducted on all raw materials, components, in-process materials, and the finished product. Along with a host of other controls (e.g., validations, calibrations, etc.), this testing assures that the final product meets its specifications and can be used in the intended manner. Information on in-process and final-product testing may be presented in the following format:

A. Introduction.
B. Outline of tests and acceptance criteria.
C. Descriptions of in-process and final-product testing methods.
D. Validation summaries.
E . Reference standards.

Lot Numbering.   The company must develop a comprehensive lot-number-ing system to enable the sponsor to track the components, raw materials, in-process product, and final product.

The sponsor must describe the lot-numbering system fully in the PLA. In the description, the sponsor must demonstrate to the FDA that the system is sufficient to track and identify the product as it advances through each phase of the manufacturing process (e.g., cell culture, purified bulk, etc.). The FDA insists on a comprehensive lot numbering system to ensure that the manufac-turer can trace the product at any stage to investigate and to identify problems.

Section 7 - Nonclinical Pharmacology and Toxicology   The PLA must describe all nonclinical pharmacology and toxicology studies conducted on the biological product. These nonclinical laboratory studies include studies submitted in the IND, studies submitted during clinical investigations, and new nonclinical studies not previously submitted. CBER reviews these stud-ies to evaluate their adequacy and comprehensiveness, and to ensure that there are no inconsistencies or inadequately characterized toxic effects.

The application should also include information on studies not performed by the sponsor, but of which the sponsor has become aware (e.g., studies in published literature).

Content requirements for the nonclinical pharmacology and toxicology section are not defined specifically in CBER regulations or guidelines. The *Draft Application for License of Recombinant DNA Derived Biological Products*, for example, states that the section should provide three elements: a nonclinical summary, toxicology reports, and pharmacology reports. In con-trast, CBER's *Application for License for the Manufacture of Interferon* states only that the PLA should include "Pre-clinical and clinical evidence of safety and efficacy of the product."

In general terms, however, the principal content requirements for the PLA's nonclinical section can be characterized as follows:

1. studies of the product's pharmacological actions in relation to its proposed indication, and studies that otherwise define the pharmacological properties of the product or that are pertinent to possible adverse effects.

2. studies of the product's toxicological effects as they relate to the product's intended clinical uses, including studies assessing the product's acute, subacute, and chronic toxicity, and studies of toxicities related to the product's particular mode of administration or conditions of use.

3. studies of the product's effects on reproduction and on the developing fetus, if necessary.

4. any studies of the product's absorption, distribution, metabolism and excretion (ADME) in animals; and

5. a statement that each nonclinical laboratory study was conducted in compliance with Good Laboratory Practice (GLP) regulations, or, if a study was not conducted in compliance with those regulations, a brief statement of the reason for noncompliance.

For each study identified above, the applicant should include a summary followed by a full report, including data and statistical analyses. Summaries help the FDA reviewer to obtain a brief analysis of the study.

Following the summaries and full reports, the applicant should provide an integrated report. Such a report integrates results from all pharmacology studies in a single, comprehensive analysis. Sponsors should provide a similar report for the toxicology information.

Section 8 - Clinical Section The PLA's clinical section is clearly the single most critical element of the application. Included in this section are the safety and effectiveness data so pivotal to the FDA's decision-making process. The clinical section, which includes the statistical component, is also likely to be an application's most complex and voluminous section.

Specifics on content requirements for the PLA's clinical section are scarce. CBER's *Draft Application for License of Recombinant DNA Derived Biological Products*, for example, identifies four specific requirements for the PLA's "Clinical Information" section: "an overall clinical summary, individual medical summaries per clinical site, FDA forms 1639, and statistical analysis per clinical site." In contrast, CBER's *Application for License for the Manufacture of Bacterial Vaccines and Antigens* states only that the PLA should include "... clinical evidence of safety, potency, and effectiveness of the product including summaries of results from all pertinent studies contained in [the] IND filing."

In general terms, however, the clinical section should consist of nine basic elements:

1. A description and analysis of each clinical pharmacology study of the biologic, including a brief comparison of the results of the human studies with the animal pharmacology and toxicology data. This section should include descriptions of whichever of the following types of studies were conducted:

   a. pilot and background studies conducted to provide a preliminary assessment of ADME as a guide in the design of early clinical trials and definitive kinetic studies;

   b. pharmacokinetic studies, descriptions of which must include a discussion of the analytical and statistical methods used in each study; and

   c. other *in vivo* studies using pharmacological or clinical endpoints.

2. A description and analysis of each controlled clinical study pertinent to the biologic's proposed use, including the protocol and a description of the statistical analyses used to evaluate the study. If the study report is an interim analysis, this must be noted and a projected completion date provided. Controlled clinical studies that

have not been analyzed in detail should be provided, including a copy of the protocol and a brief description of the results and status of the study.

3. A description of each uncontrolled clinical study, a summary of the results, and a brief statement explaining why the study is classified as uncontrolled.

4. A description and analysis of any other data or information relevant to an evaluation of the product's safety and effectiveness obtained by the applicant from any foreign or domestic source.

5. An integrated summary of the data providing substantial evidence of effectiveness for the clinical indications. Evidence is also required to support the dosage and administration section of the labeling, including support for the dosage and dose interval recommended and modifications for specific subgroups of patients.

6. A summary and update of safety information.

7. An integrated summary of benefits and risks of the biologic, including a discussion of why the benefits exceed the risks under the conditions stated in the labeling.

8. A statement noting that each human clinical study was conducted in compliance with the IRB regulations and with the informed consent regulations. If the study was not conducted according to these regulations, the applicant must state this fact and the reasons for non-compliance.

9. Statistical information with which FDA reviewers can assess the validity of key evidence supporting a biologic's safety and effectiveness. This section should contain:

   a. information concerning the analysis of each controlled clinical study, and the documentation and supporting statistical analysis used in the evaluation of the controlled clinical studies establishing effectiveness; and

165

      b. a summary of information on the safety of the biological product, and the documentation and supporting statistical analysis used in evaluating the safety data.

    Applicants are encouraged to meet with CBER statisticians prior to preparing the clinical section of the PLA. Ideally, such meetings produce sponsor-CBER agreements on the format and tabulation of clinical data and the statistical approaches used to analyze the clinical results.

As in the preclinical section, each individual clinical study should have a summary report followed by a full report with data and statistics. Typically, the general format is as follows:

Phase 1 Studies

- summary of each study;
- full report with data and statistics of each study; and
- integrated report for all studies.

The same information is necessary for Phase 2 and 3 studies. If possible, the sponsor should develop an overall integrated report for the three phases of clinical investigation.

Section 9 - Product Stability Section In this section, the applicant must present information and data demonstrating the stability of both the biological substance and the final product. This section should contain the following:

- a summary of the stability data for the biological substance and final product;
- the protocol for the stability testing conducted;
- real-time stability data for the biological substance and final product followed by a full discussion of the data;
- accelerated data, if any such studies were conducted;
- container/closure integrity studies;

- information on expiration dating and storage conditions, including how these relate to the stability data generated; and
- standard operating procedures for the product's sample retention program.

Section 10 - Facilities: Systems and Design   Although the ELA provides detailed facilities-related information, the PLA must provide product license reviewers with an overview of the manufacturing facility and its operations regarding the product.

The applicant's goal in this section is to provide the PLA reviewer with a complete description of the entire manufacturing operation, from facilities to product. This section should consist of the following principal elements:

- a diagram of the entire building, accompanied by a narrative identifying the materials used in the building's construction and the operations performed in the building;
- a diagram of the manufacturing facility, accompanied by a narrative identifying the materials used in the facility's construction and the operations performed in the manufacturing area;
- an identification of each room in the manufacturing area, accompanied by a narrative on each room, including construction materials and special systems (e.g., air and water systems);
- a listing of all operations performed in each room;
- a listing of equipment located in each room;
- a diagram of the facility's water system and the drop points for each room, accompanied by a narrative that describes the type of water system, the composition of materials used in the system, and the quality of water produced, and that specifies whether the system is a circulating or non-circulating loop system, whether the system is loop-heated (or is not heated), and whether water is dumped or stored (and if so, how it is stored);
- a diagram on each room's air system, accompanied by a narrative describing the type of air system, the quality of air produced, the type of filters used, the quality of air provided to each room, air changes in each room, and pressure differentials between rooms;

- a diagram and narrative describing the flow of personnel through the manufacturing facility; and
- a diagram and narrative regarding the flow of materials through the manufacturing facility.

The flow of personnel and materials should demonstrate that clean products, materials, equipment, and personnel never come into contact with dirty products, materials, equipment and personnel, and that there is adequate separation between clean and dirty entities in distance or time. This will address FDA concerns regarding the potential for product contamination.

Section 11 - Labeling Section   The PLA must include a section providing draft labeling for the container label, package label, and package insert. These three elements of product labeling should: summarize the essential scientific information needed for the safe and effective use of the biologic; be informative and accurate and not promotional, false, or misleading; and be based, whenever possible, on data obtained from human use of the product.

Just prior to approval, this section will be updated to include final printed labeling and any advertising and promotional materials to be used at product launch. At the time of the PLA submission, however, the sponsor must provide draft labeling that consists of the following three elements:

1. draft labeling for the container label, which must include the following information:

   - the product's proper and trade names;
   - the product's lot number;
   - the product's expiration date;
   - the product's approved dose;
   - a statement that federal law prohibits dispensing the product without a prescription;
   - the manufacturer's name and address;
   - the manufacturer's license number; and
   - the product's National Drug Code (NDC) number.

If the container label cannot accommodate all these elements, then the proper name, lot number, and manufacturer's name and address must be placed on the partial label. If the product container cannot accommodate any labeling, all labeling is then included on the package label.

2. draft labeling for the package label, which must include the following information:

- the product's proper and trade names;
- the manufacturer's name and address;
- the product's lot number;
- the product's expiration date;
- the names of preservatives used in the product;
- the number of containers in the package;
- the package's contents;
- recommended storage conditions;
- special instructions (e.g., shake well);
- the product's dose and route of administration;
- the names of any antibiotics used during manufacturing;
- the product's source;
- a statement that federal law prohibits dispensing the product without a prescription;
- a reference to the package insert; and
- the product's NDC number.

3. draft labeling for the package insert, which should include the following sections:

- product description;
- clinical pharmacology;
- indications and usage;
- contraindications;
- warnings;
- precautions;
- adverse reactions;
- drug abuse;
- overdose;
- dosage and administration; and
- drug supply (i.e., dosage form, strength, unit availability, etc.).

Section 12 - Product Samples   Before a biological product is marketed, the FDA will want to validate the sponsor's characterization methods for both the biological substance and the finished product. Therefore, at some point during the PLA/ELA review process, the FDA will request three items: a biological substance sample, a finished product sample, and the sponsor's reference standards. The applicant should forward sufficient sample quantities to CBER to allow the center to duplicate all testing requirements associated with the biological substance and finished product.

Usually, applicants provide three lots of the finished product as samples. Ideally, all three samples will be taken from lots used during pivotal studies. At a minimum, however, one sample should be derived from a lot used in pivotal studies. Once the PLA and ELA are approved and these lots are certified by the FDA, the product can be marketed.

As of this writing, CBER had not issued a definitive policy on when product samples should be submitted during the PLA/ELA review process. However, because of the aggressive review timelines imposed on CBER under the Prescription Drug User Fee Act of 1992, some CBER officials believe that product samples may soon have to be submitted with the PLA and ELA. In the meantime, applicants should seek CBER guidance regarding the submission of product samples.

Section 13 - Responsible Personnel   In the PLA, applicants must identify and provide relevant information on at least four individuals considered "responsible personnel":

- the head of quality assurance;
- the head of quality control;
- the head of manufacturing; and
- the responsible head.

At a minimum, this section should provide résumés for each of these individuals. In addition, the applicant should include an organizational chart sufficiently detailed to include the heads of manufacturing and quality functions.

Section 14 - Environmental Assessment   The National Environmental Policy Act (NEPA) requires that each PLA submission include an environmental assessment report. NEPA is the national charter for the protection, restoration, and enhancement of the environment. The environmental assessment report assures the FDA and the Council on Environmental Quality that all sources of materials entering the environment have been identified and are controlled by the applicant's manufacturing process.

Federal regulations (21 CFR 25.31a) specify the format for the environmental assessment report (see Chapter 9). Basically, the assessment requires that the applicant specify how its manufacturing processes affect the air, the water supply, and waste disposal systems. The applicant must also certify that these processes meet all federal, state and local emission requirements.

Section 15 - Certification of Regulatory Compliance   As part of any PLA filing, the applicant must certify compliance with applicable FDA regulations. By signing a certification statement on the PLA application form, or a similar statement in the PLA cover letter or elsewhere in the application, the responsible head certifies that all statements made in the filing are "true and correct," and that the sponsor complied with applicable federal regulations, including GLP, cGMP, and GCP.

All PLAs submitted on or after June 1, 1992 must also certify that the applicant did not and will not use the services of individuals or firms that have been debarred by the FDA. Under the Generic Drug Enforcement Act of 1992, the FDA is authorized to debar individuals convicted of crimes relating to the development, approval, or regulation of drugs or biologics from providing any services to applicants. The statute states that applications for drug products (including biological products) must include "a certification that the applicant did not and will not use in any capacity the services of any person debarred ... in connection with such application."

Section 16 - Other Information   The sponsor may submit in this section any other information that may help the FDA to evaluate the safety and effectiveness of the product.

Assembling and Submitting the PLA
The PLA should be properly indexed and paginated for ease of review. Each

volume should be no more than two inches thick, bound on the left side of the page, and printed on standard U.S. paper (8.5 x 11"). The front cover of each volume should specify the name of the applicant, the name of the product, and the PLA number (if known).

The lower right hand corner of each volume should read: This submission: "Vol. ___ of ___ vols." The upper right of each volume should read: "Volume ___."

Applicants must submit three copies of the PLA (and ELA) to CBER. The copies may be hand-delivered or mailed. If the applicant hand-delivers the PLA, the sponsor should bring an extra copy of the cover letter with the shipment so that the letter may be date-stamped upon delivery to the FDA's Document Control Center. This letter provides evidence that the document was submitted to the FDA.

If forwarded by standard mail service, the PLA shipment should include a letter of instructions with a document stating that the PLA submission has been received by the FDA. The FDA document control person will sign the document, place it in a stamped return envelope provided by the sponsor, and return it to the company. This document also serves as proof that the PLA was submitted.

Sponsors mailing the PLA should forward the application and all related submissions to the following address:

> Center for Biologics Evaluation and Research
> Food and Drug Administration
> 1401 Rockville Pike
> Suite 200N, HFM-99
> Rockville, MD 20852-1448

If the applicant forwards the PLA using a commercial overnight service, a return receipt is provided. In this case, the receipt provides evidence of the PLA's delivery.

Amending the PLA

During the review of the PLA, the FDA is likely to request additional information to address unresolved issues concerning the original submission. A response to such a request is generally referred to as a PLA amendment.

The content of a PLA amendment will, of course, depend exclusively on the nature of CBER's information request. The format used in submitting such amendments is similar to that used for the original PLA submission. The cover letter for the amendment should be titled:

"Amendment to PLA _____"

In the cover letter, the applicant should clearly identify the purpose of the amendment and the contents of the submission. The amendment should be paginated in a manner that will allow CBER to locate the section of the PLA in which the amendment should be incorporated.

Supplements to the Original PLA

A holder of an approved PLA may seek to change its manufacturing methods, expand the product's indication, or make other changes that reflect new technology or make their product or processes more competitive. In such cases, the company must submit a supplemental PLA. While PLA amendments are submitted to update or modify an unapproved PLA, supplements are submitted to modify approved PLAs.

Federal regulations state that all changes in manufacturing methods and labeling and important changes in location, equipment, and management and responsible personnel may not become effective without prior FDA approval. Since such changes can be expected throughout a product's lifecycle, supplementing the PLA becomes an ongoing process. Supplement-related activity is particularly high as a company refines, scales-up, and streamlines its manufacturing operation.

Content requirements for supplemental PLAs will depend on the nature of the proposed change. In general terms, the applicant must provide new data and information sufficient to support the modification. In filing a supplement to the PLA, sponsors should follow the same instructions as those provided above for amending the PLA.

Compared to companies holding approved NDAs for drug products, biologic licensees have considerably less latitude in making minor changes to labeling or manufacturing processes without first obtaining FDA approval through a supplemental application. Although there were ongoing efforts within CBER to provide more flexibility to PLA/ELA holders, no new policies had been finalized as of this writing.

Chapter 9:

The Establishment License Application and Current Biological Manufacturing Arrangements

by Thomas R. Monahan, Ph.D.
Manager, Biological Quality Control
Lederle-Praxis Biologicals

To market a biological product legally, the Public Health Service Act requires a manufacturer to apply for and obtain two types of licenses from the FDA. The first of these, a product license, permits a manufacturer to produce a biological product for commercial purposes (see Chapter 8). The second, an establishment license, permits a specific facility to manufacture the product.

Unlike the PLA, whose function is to provide evidence that a new biological product meets safety, effectiveness, potency, purity, and other standards, the ELA's function is to provide evidence that a specific manufacturing facility is capable of producing a biologic according to current good manufacturing practices, relevant federal regulations, and standards specified in the PLA. Therefore, the ELA supplies information about a facility's physical design and construction, utilities and systems, equipment, production activities, personnel flow, control methods, and other key functions and characteristics.

The Establishment License Application: An Overview

In general, an ELA consists of a single nine-page form, *Application for Establishment License for Manufacture of Biological Products* (FDA Form 3210), which contains detailed information about the design and operation of the facility(ies), systems, equipment, and personnel involved in the manufacture of a particular biological product. It is important to note here that the FDA defines manufacture as "all steps in propagation or manufacture and preparation of products and includes but is not limited to filling, testing, labeling, packaging, and storage by the manufacturer" (1).

Although this chapter focuses on ELA Form 3210, it is worth noting here that licensed blood banks have a special ELA form. Blood banks requiring licensure—in other words, those collecting or distributing blood on an interstate basis—must file Form 2599, *Establishment License Application for the Manufacture of Blood and Blood Components*.

Typically, an ELA is submitted with the PLA for the subject product. Assuming that the review reaches this stage, the FDA must approve the applications simultaneously.

When a company files an ELA and is initially granted an establishment license, the FDA assigns the manufacturer a unique license number, which must appear on product labeling. Interestingly, if the company is granted licenses for other biological products and manufacturing sites, the original license number will apply to those products and sites as well. In other words, the FDA will not assign a second establishment number to the manufacturer.

When a manufacturer seeks to market and produce a different biological product than that covered in the original ELA, the company must amend the existing ELA to reflect all new facilities, equipment, services, and personnel involved in the production of the new product. In this way, all subsequent ELAs submitted by the manufacturer are simply amendments to the company's original ELA approved by the agency.

Although the ELA is unique to biological products, it is similar to a master file in many respects, primarily in that both documents contain information relevant to the manufacture of a product. There are notable differences between a master file and an ELA, however. The most significant difference is that a master file is normally reviewed by the FDA as supporting documentation to a premarketing submission and is not formally approved. On the other hand, ELAs are considered "stand-alone" documents that are thoroughly evaluated for the

176

purpose of approval, often by the same CBER reviewers who comprise the review committee for the corresponding PLA. Also, while the format of a master file is specified in FDA guidelines, an ELA must be submitted on a specific FDA-issued form—Form FDA 3210 (2).

Frequently, the similarities between master files and ELAs lead to confusion regarding when each is appropriate. Although the utility of master files is limited largely to contract manufacturing arrangements, a master file often is submitted when an ELA is required. The confusion may be particularly acute in cooperative manufacturing arrangements such as shared and divided manufacturing scenarios, in which the principal participants must submit ELAs (see discussion below).

The ELA: An Analysis of Content Requirements
While the FDA has detailed regulations and more than a dozen guidelines identifying submission requirements for new drug applications (NDA), the agency provides relatively little formal guidance on the form and content of PLA and ELA submissions. However, CBER has always publicly maintained that it will provide information and guidance, either formally or informally, to any potential biological product manufacturer making such a request. In addition, CBER personnel have been active participants in industry and trade association symposia and seminars, and have used these forums to disseminate and discuss information regarding submissions.

As discussed above, ELAs are submitted on Form FDA 3210, which outlines the content requirements for the filing (see copy of Form FDA 3210 at the conclusion of this chapter). And although the latest version of Form FDA 3210 has an expiration date of May 31, 1992, CBER accepts this iteration of the form. CBER recently announced that a revised ELA form is being prepared, although the center has no timetable for its issuance.

The nine-page ELA form identifies 11 content sections that applicants must address in the submission (author's note: due to an error on Form FDA 3210, the "Section VII" heading is used for two separate sections):

I.   General Information
II.   Building and Facilities
III. Animal Facilities
IV.   Work With Microorganisms
V.   Equipment

Section I. General Information   This section of the ELA must provide the name, address and telephone number of the manufacturer applying for the license as well as similar information for each location at which manufacturing will take place. Each location must also be identified as to the type of manufacturing (i.e., production, testing, labeling, storage, etc.) to be performed at the site. If in the production of the final product the manufacturer obtains source material from another source, demographic information about this additional site(s) must also be provided.

Finally, this section must provide the name and address of the "responsible head" or the responsible officials to whom copies of official correspondence should be sent. The responsible head is a position unique to a U.S. biological product manufacturer. Regulations require this person to, among other duties, "exercise control of the establishment in all matters relating to compliance with the provisions of applicable federal regulations and with commitments made in an ELA and PLA" (3). Therefore, the responsible head represents the biological manufacturer in all licensing and compliance activities.

Section II. Building and Facilities   Sponsors should treat the description of the building(s) and facilities associated with the manufacture of a biological product as a critical element of an ELA. In this section, a manufacturer must describe in detail all areas directly involved in the manufacturing process. Given the level of detail typically required, the section should provide a description of the design and construction features of each relevant manufacturing area, including the composition of all wall, ceiling, and floor surfaces. The section also must describe the functions performed in each area.

The applicant must identify areas of sterile operations. If the facility performs multiple sterile processes in these areas, the sponsor must describe the procedures in place to monitor and prevent material or product contamination.

178

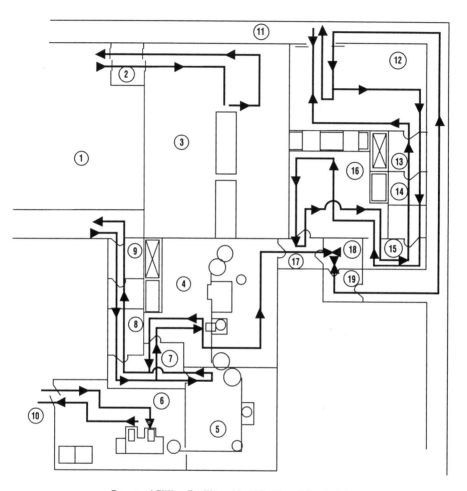

Proposed Filling Facility—Identification of Product Areas

1. Warehouse
2. Airlock
3. Vial Washing Depyrogenating
4. Filling /Stopping
5. Capping
6. Trayloading
7. Airlock
8. Gown Room
9. Airlock
10. Shipping
11. Corridor

12. Stopper/Parts Washing
    and Preparation
13. Airlock
14. Gown Room
15. Airlock
16. Stopper/Parts Unloading
    Sterile Side
17. Airlock
18. Bulk Fill/Parts
    Transfer Room
19. Airlock

Legend:

——— Personnel and
    Equipment flow

179

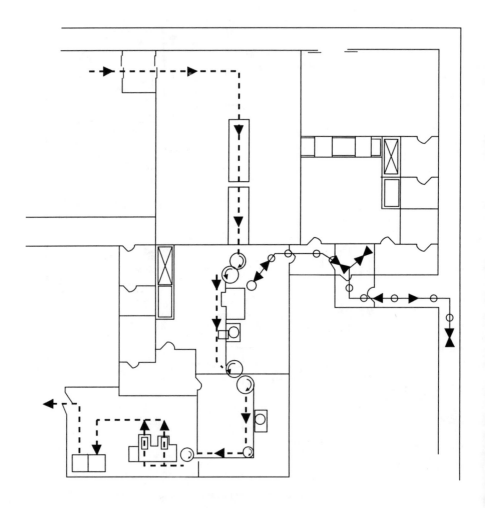

Proposed Filling Facility—Material Flow

Legend:

– – – – – – Glass Vial Routing

o o o o o Bulk Material Routing

The ELA also requests a description of the flow of personnel to and from each area. Typically, this information is best presented by providing a schematic diagram of the product area(s) (see first diagram above). Superimposed on the diagram would be not only personnel flow but also the flow of raw materials and finished product (see second diagram above). Such a diagram can be used to address the ELA submission requirements outlined in Section VIII.

Information about relevant air and water systems that service these manufacturing areas is required. The sponsor must provide specifications for each area of use and a description of the validation of each system. An outline of pest control procedures as well as relevant building maintenance schedules must be supplied. Finally, this section must include information about the relationship between the manufacturing facilities and nonmanufacturing portions of that same facility and surrounding buildings.

If the proposed manufacturing facility is part of a larger complex, the submission of a general facility map indicating the location of the manufacturing facility is extremely helpful. A brief description of all surrounding industry, farms and residential property is required.

Section III. Animal Facilities   Animals are often an integral component of a biological product's manufacture, especially in the testing of these products. Many of the tests described in federal regulations for biological products use a variety of animal species. Therefore, information about the types of animals used in the production and testing of a proposed product is required in the ELA.

This section must specify the location and construction features of all animal facilities as well as all procedures for animal quarantine and husbandry. If specific animal testing is performed at another manufacturing site or by a contract testing laboratory, a description of this arrangement must be included in this section of the ELA.

Section IV. Work with Microorganisms   The production of many biological products, particularly vaccines and innovative products manufactured using recombinant DNA (rDNA) techniques, involves the use of various species of bacteria, fungi, and viruses. Certain testing procedures employ microorganisms as well.

181

An ELA must identify any microorganisms stored in an establishment, and the purpose(s) for which these are used. The application must specify all precautions employed in the storage and handling of all microorganisms used in the manufacturing process.

CBER has expressed special concern about the use of spore-forming microorganisms in manufacturing processes, a concern that is reflected in federal regulations (4). The ELA must identify the microorganisms and the conditions under which they are stored and must provide a complete description of the use of such microorganisms in the manufacturing process. Often, spore-forming microorganisms are used in the validation and testing of various sterilization procedures. It is important to note that the FDA allows "Spore-bearing organisms (to be) used as an additional control in sterilization procedures (and to) be introduced into areas used for the manufacture of products, only for the purpose of the test and only immediately before use for such purposes" provided certain conditions are met (5).

Section V. Equipment   The fifth section of an ELA should provide a listing and discussion of the major equipment used in the manufacture of the proposed biological product. The sponsor must identify each major piece of equipment, its manufacturer, and its location in the manufacturing facility(ies). One option for addressing these requirements is to include a tabular presentation referenced to a schematic diagram of manufacturing areas and equipment locations.

The application must describe sterilization processes for major equipment—whether autoclave, dry heat, or other methods. The sponsor must provide data to support proposed conditions and parameters for the particular sterilization method chosen.

All incubators, refrigerators, freezers and other equipment used to provide a specific type of controlled environment for raw materials, in-process materials, or final product must be identified. The ELA must specify the sponsor's methods of temperature recording and the frequency of temperature calibration. Methods and frequency of cleaning and calibration of all equipment listed must be provided.

Perhaps the most critical information required in this ELA section is a description of the methods used to clean and validate equipment involved in the manufacturing process. All major equipment used must be validated, a process the FDA has described as "establishing documented evidence which provides a high degree of assurance that a specific process will consistently produce a product meeting its pre-determined specifications and quality attributes" (6). Normally, an outline of the validation protocols and a summary of the validation data will be satisfactory for an ELA submission. However, it is always prudent to discuss such an approach with CBER prior to an ELA filing.

Section VI. Production and Testing Contrary to what is implied, the production and testing section of an ELA does not deal specifically with all the parameters involved in the manufacture of a biological product. Such detailed information normally is provided in an accompanying PLA.

Rather, in this section of an ELA, the FDA attempts to obtain information about what other products are produced or tested on the premises and if and when any diagnostic work is performed. Because the FDA does not specifically define the term "products," the sponsor must consider identifying any biological, drug, or device produced at the location. If, in fact, other products are being simultaneously or subsequently manufactured, the application must describe the methods used to prevent contamination during production and errors during product labeling.

In this section, the sponsor must identify areas in which more than one product is manufactured. Steps taken to prevent product contamination during the filling operation must be provided. The agency specifically requests information regarding sanitation, validation, and air quality designations. The ELA must identify the location of all relevant storage and quarantine areas for unlabeled and finished product as well as procedures for the storage and handling of this material. Finally, this section should describe the procedure used for the selection and storage of product retention samples.

Section VII. Records The records section of the ELA must provide a description and sample of the records related to the manufacture of a lot or batch of the proposed biological product as well as the retention period for such records. These records should be designed such that each step in the production and testing of a product lot can be verified by checking the dated signature of a responsible individual. Applications for biological products that require sterilization during manufacture must include a description and sample of the sterilization records.

A description and sample of all records that will be maintained relevant to the distribution of the final product must be submitted as part of the ELA. The applications must include an outline of the maintenance of all production records to convince CBER that a successful product recall is possible. It must provide an explanation of the lot number designation used to identify each product lot. The sponsor must specify the significance of each portion of the lot designation that appears on the final container.

An important element of this ELA section requires the sponsor to address how production, testing, and distribution records are integrated so that, starting with the lot number appearing on a final container, one can trace the manufacturing history of a particular product lot. An outline of this information flow must be provided in the application. CBER personnel may require a review and demonstration of this process during the routine pre-licensure inspection prior to the approval of a PLA and ELA.

Finally, a listing of all biological products manufactured and the approximate number of lots produced during the preceding 12 months must be provided. Information regarding licensable biological products manufactured at another site but sold by the establishment, or any licensable products partially or completely manufactured by the establishment but sold to another establishment for further manufacturing, must be listed and the conditions of this transfer must be described. A listing of any product intermediates exported is necessary.

Section VII. Organization and Personnel   According to current U.S. regulations, all individuals engaged in the manufacture of a biological product must have demonstrated capabilities commensurate with their assigned functions and a thorough understanding of the manufacturing operations they perform (7). Each person must also possess the necessary training and experience related not only to the manufacture of a particular product, but also to the applicable federal standards and guidelines relevant to such a product.

Therefore, an ELA must include an organizational chart consisting of names and titles of key personnel and applicable consultants. Defining key personnel is left to the discretion of the organization submitting the ELA, but this definition may be challenged by CBER if an abbreviated list is submitted. A curriculum vitae (CV) or some other documented evidence of the qualification and training of each person listed in the organizational chart also is required.

Section VIII. Additional Information   This section of the ELA has three parts, only two of which need be addressed if the application being submitted is an amendment to an initially approved ELA. If the ELA being submitted is an initial submission, the third element—a copy of the sponsor's Certificate of Incorporation—must be included.

Copies of the relevant floor plans and the facility layout must be submitted to complete this part of the ELA. This information usually is presented in the form of schematic diagrams, which should be annotated as to the dimensions of the areas shown.

The ELA must specify the location of major equipment and lavatory and dressing room facilities. Placement of air handling and water systems must be identified. The facility layout must indicate the flow of raw materials, finished product, and personnel throughout the facility. The information and diagrams presented in this section may be used to satisfy some of the requirements of ELA Section II. Building and Facilities, which may be cross-referenced.

An environmental assessment, or EA, (see exhibit below) has become an important requirement for drug and biological approval in the United States (8). The EA is a public document in which sponsors submit environmental and other pertinent information regarding a proposed action—in the case of a PLA and ELA, the action is the manufacture of the biological product. The FDA reviews this information to determine whether an Environmental Impact Statement (EIS) or a Finding of No Significant Impact (FONSI) is necessary.

The level of detail necessary in EAs for drugs and biological products manufactured using techniques of biotechnology has been the subject of public discussion between the FDA and regulated industry (9). Clearly, the FDA will now closely scrutinize the information and data contained in an EA, and, in some cases, will require very specific and detailed information prior to the approval of either a new drug application or PLA/ELA submission.

Section IX. Comments The "Comments" section of an ELA gives the sponsor an opportunity to present any information that it believes would be useful in supporting the application. The comments may address specific questions or concerns previously raised by CBER relevant to the suitability of the manufacturing establishment.

Because each ELA is product-specific, any comments included will most likely apply to a particular establishment. Therefore, guidance on the contents of this section should be given only on a case-by-case basis.

Environmental Assessment

For proposed actions to approve food or color additives, drugs, biological products, animal drugs, and class III medical devices, and to affirm food substances as generally recognized as safe (GRAS), the applicant or petitioner shall prepare an environmental assessment in the following format:

1. Date.
2. Name of applicant/petitioner.
3. Address.
4. Description of the proposed action.
5. Identification of chemical substances that are the subjects of the proposed action.
6. Introduction of substances into the environment.
7. Fate of emitted substances in the environment.
8. Environmental effects of released substances.
9. Use of resources and energy.
10. Mitigation measures.
11. Alternatives to the proposed action.
12. List of preparers.
13. Certification.
14. References.
15. Appendices.

Source: 21 CFR 25.31 (a)

Section X. Names and Titles of Experts Responsible for the Production and Testing of Product As the title of this section states, the names, titles, and signatures of individuals responsible for the production and testing of the subject biological product must appear on the completed ELA.

Normally, this would include production, quality control, and quality assurance personnel. The completed ELA Form FDA 3210 must include the dated signature of the responsible head.

The Submission of the ELA

The biological sponsor must submit a "completed" ELA before CBER takes action on the application. Current regulations state that an application for license—PLA or ELA—will not be considered to be filed until all pertinent information and data have been submitted to CBER (10). Normally, the manufacturer's responsible head must file three copies of a completed ELA to CBER's director. *(Editor's note: For a complete analysis of CBER's ELA review process, see chapter 10)*

Unlike CDER, which asks drug sponsors to submit "archival" and "review" copies of NDAs, CBER makes no such distinctions between the three ELA copies filed with the center. Also, CBER generally discourages sponsors from submitting "desk copies" of ELA submissions directly to reviewers because the center cannot ensure the confidentiality of these submissions. However, under certain circumstances discussed with CBER in advance, sponsors may send certain ELA information directly to a reviewer as a "desk copy."

Post-Approval Considerations and the ELA

As with other regulatory submissions, the FDA and the sponsor have ELA-related obligations following a facility's licensure. Licensees must report to the FDA significant manufacturing-related changes, and the FDA must periodically inspect manufacturing facilities and employ ELA suspension and revocation powers when necessary.

ELA Amendments Many of the biological products currently marketed in the United States have, for the most part, been manufactured for some time. More recently, novel therapeutic biological products have been developed and manufactured using techniques of biotechnology that were not available several years ago.

It is unreasonable to expect that facilities initially described in an ELA for production and testing of the earliest vaccine products would not have undergone changes since their introduction. Neither is it plausible to believe that many of the state-of-the-art production facilities and equipment employed in the manufacture of today's biological therapeutics will survive change, especially as technology continues to evolve at a rapid pace. Therefore, changes in the manufacturing process, responsible personnel, facilities, and equipment are inevitable.

Before making "important" changes to any facilities, services, equipment, or personnel involved in the manufacture of a biological product, sponsors must report and obtain CBER approval for the changes through ELA amendments (11). CBER has always contended that products can be made with these changes prior to agency approval but cannot be released for sale until such changes are approved.

Unfortunately, CBER has yet to specifically define "important" manufacturing changes. This has caused a running debate between manufacturers and CBER as to what specific changes must be reported and approved in ELA amendments prior to implementation. Recently, several trade organizations representing U.S. biological product manufacturers have initiated a dialogue with CBER management in an attempt to clarify these and several other key issues related to the manufacture of biological products. Industry hopes that CBER will reduce reporting requirements and provide more flexibility regarding manufacturing changes.

The FDA has not specified a standard format for ELA amendment submissions. CBER has indicated that sponsors should communicate with the center prior to the filing of an amendment.

Existing facilities seeking to add new products and facilities should, in the ELA amendment, provide detailed discussions of areas that will be the subject of change and areas that will remain unaffected. This may require referencing previously submitted (and approved) information. To facilitate and expedite the review process, however, the sponsor may choose to resubmit this information with the ELA amendment.

Facility Inspections   After an establishment license has been issued, a biological manufacturing facility is subject to periodic inspection. Each licensed establishment and any additional locations engaged in the manufacture of biological products must be inspected by the FDA at least once every two years (12).

For various reasons, CBER may choose to inspect an establishment more often. More frequent inspections are particularly likely for establishments engaged in the manufacture of several different biological products and for manufacturers or products that have been the subject of recent FDA compliance actions.

Periodic biological establishment inspections typically are unannounced. However, when a manufacturing facility is located outside the United States or there exists a specific desire for agency personnel to observe a particular phase of a manufacturing process, CBER typically will notify an establishment prior to the inspection.

Like pre-licensure inspections, periodic biologic inspections provide an opportunity for trained CBER personnel to observe and review various aspects of a manufacturing operation and the relevant facilities in which these operations are performed. The goal of the inspection is to ensure continued compliance with parameters defined in existing PLAs, ELAs, and applicable federal regulations. The inspections also allow CBER personnel to evaluate relevant production and testing records for products previously manufactured and distributed (13).

Historically, periodic inspections have been the purview of CBER personnel, particularly members of the research staff. However, due to the large number of blood banks and blood banking establishments, responsibility for the inspection of these facilities was transferred to FDA field inspectors. It is highly probable that the responsibility for all annual biological inspections may ultimately be transferred to FDA field inspectors. This is likely to follow an adequate period of training which will, no doubt, be characterized by joint inspections conducted by CBER personnel and FDA field staff.

Joint inspections of biological manufacturing facilities are, in fact, becoming more commonplace. Although these are designed to leverage the expertise and experience of FDA field inspectors, CBER personnel will continue to take the lead role in joint inspections.

ELA Suspension and Revocation

An ELA and PLA will remain valid unless suspended or revoked by CBER (14). License suspensions and revocations are designed to prevent a manufacturer from producing and distributing a product that has been found to be either unsafe or ineffective. Several circumstances may lead to such actions (15):

- The manufacturer may voluntarily have a license revoked after notifying CBER of its desire to discontinue the manufacture of a licensed product (16).

- The licensed product is not safe and effective for all of its intended uses, or is misbranded with respect to any such uses.

- The FDA cannot gain access to an establishment to conduct a meaningful inspection.

- The manufacture of a product(s) has been discontinued to an extent that a meaningful inspection or evaluation cannot be made.

- The manufacturer has failed to report an important change in manufacturing or facilities.

- The establishment or product fails to conform to the applicable standards described in the license or required by federal regulation.

- The establishment or the manufacturing procedures have been changed to a degree requiring a demonstration that the modifications meet federal regulations.

While each is designed to prevent the production and distribution of an unsafe or ineffective product, there are at least two important differences between license suspensions and revocations. Typically, a license suspension would be the FDA's first-line response to a product safety or effectiveness issue. If the condition that precipitated the suspension is not corrected to the FDA's satisfaction, a license revocation may follow.

Unlike a suspended license, which can be reinstated if the situation(s) that prompted the suspension is corrected by the manufacturer, a revoked license may not be reinstated. The sponsor of a revoked license must reapply and be granted a new license before manufacturing and marketing the subject product.

Chapter 9: The Establishment License Application and Current Biological Manufacturing Arrangements

If the FDA suspends a license, a manufacturer must notify those who sell or distribute the product of the suspension and provide CBER with a copy of this notification and a listing of those notified (17). The manufacturer may attempt to resolve the situation by requesting a hearing with CBER. If granted, such a hearing should proceed on an expedited basis (18). Notification of a license revocation and the cause for such an action must be announced in a *Federal Register* notice (19).

Current Manufacturing Arrangements For Biological Products

The manufacture of biological products in the United States has, historically, been undertaken by individual commercial manufacturers carrying out all aspects of the complex manufacturing process. But new biological technologies and the resulting products have spawned a new biotechnology industry comprising not only large multinational pharmaceutical companies but also many "start-up" companies founded to commercialize a single compound or technology. Such new companies are quite different from the traditional pharmaceutical giants in that their entire existence, in some cases, is directly dependent on the clinical success of a single compound.

It is virtually impossible for such new companies to possess all the resources needed to develop and commercialize a pharmaceutical product in the United States. The nature and costs of modern biological product development have put companies under considerable economic pressure to forge relationships with other companies to develop and manufacture products (20). As a result, biotechnology methodologies have created new and complex biological manufacturing arrangements not typically encountered by CBER (21).

To date, CBER has recognized and allowed the formation of certain joint manufacturing arrangements or alliances for biological products. The number of these arrangements between companies engaged in the manufacture of biological products certainly will increase. Furthermore, novel manufacturing arrangements will, no doubt, be proposed by manufacturers as they attempt to commercialize existing scientific technologies (22).

Currently, CBER reviews all proposed joint manufacturing arrangements on a case-by-case basis. Formal guidance on these arrangements is scarce, at best. In November 1992, however, CBER issued a policy statement, *Manufacturing*

Arrangements for Licensed Biologicals, to clarify its position on these "cooperative" manufacturing arrangements (23).

The policy statement is designed to provide guidance, not establish requirements or standards, for manufacturing arrangements. Furthermore, CBER cautions that the policy statement is not comprehensive, that it may not be applicable to all situations, and that sponsors may follow alternative procedures provided they meet with the center's approval.

In federal regulations and its new policy statement, CBER recognizes several different forms of biologic manufacturing, including:

• sole manufacturing
• shared manufacturing
• contract manufacturing
• divided manufacturing

Sole Manufacturing

The predominant mode of biological product manufacturing in the United States involves a manufacturer operating at a single site and performing every step in a manufacturing process. This type of manufacturing arrangement is called sole manufacturing.

Many biologic manufacturers are involved in the manufacture of several biological products. These establishments operate under a single establishment license number and hold various product licenses for the products they manufacture. Different biological products generally do not use the same areas and equipment for production purposes. However, they often share other manufacturing areas such as testing laboratories, sterilization equipment, and filling and packaging facilities.

When faced with a short product supply, a licensed biological manufacturer may obtain and use certain source material manufactured at an unlicensed facility. This arrangement is possible only under the following conditions: (1) manufacturing at the unlicensed facility will be limited to the initial and partial manufacturing of a product for shipment solely to the licensee; (2) the unlicensed manufacturer is registered with the FDA in accordance with registration and listing requirements; (3) the product made at the unlicensed facility is in short supply due either to peculiar growth requirements or scarcity of the source organism required for manufacturing; and (4) the licensed manufacturer can assure that, through inspections, testing, or other arrangements, the product made at the unlicensed facility will be made in full compliance with applicable regulations. Due to these conditions, short supply arrangements have fairly limited applicability.

Historically, biological manufacturers have neither designed nor operated multi-use facilities in which two or more products are processed in the same area, either concurrently or on a campaigned basis. Recently, however, manufacturers responsible for the sole manufacture of several biological products have begun to explore the feasibility of employing multi-use facilities in the manufacture of biological products. This initiative is a symptom of both the technical advances in production methodologies and analytical techniques and the economic realities of today's pharmaceutical industry. A collaborative industry/ FDA effort focused on establishing realistic guidelines for the manufacture of biological products in multi-use facilities is the obvious goal.

Shared Manufacturing

As stated above, the traditional concept of a single or sole manufacturer completing every step of a complex manufacturing process has been greatly affected by the biotechnology industry. In an attempt to deal with the needs and limitations of this fledgling industry, CBER has allowed "shared manufacturing" arrangements for biological products.

The shared manufacturing concept has a relatively short history. It was not until 1983, when the FDA issued a *Federal Register* notice concerning the licensing of a monoclonal antibody product prepared by hybridoma technology, that shared manufacturing arrangements were openly acknowledged (24).

Shared manufacturing is an arrangement in which two or more firms agree to transfer manufacturing responsibilities at a predetermined step. In typical shared manufacturing arrangements, one firm takes product processing to a specific stage, at which point a second firm assumes responsibility for the product's completion. The participating companies, often small biotechnology firms, may have neither the desire nor the capability to complete every step in the manufacturing process.

Essentially, a shared manufacturing arrangement allows two or more manufacturers to participate, or "share," in the manufacture of a biological product through the issuance of restricted product licenses. Each manufacturer must submit a PLA for its portion of the manufacture. If the FDA approves, the initial manufacturer receives a restricted product license for initial product processing, while the second manufacturer receives a PLA for the subsequent manufacturing steps.

In addition to a restricted product license, all participating manufacturers must either be issued an establishment license or amend an existing establishment license. If any changes are made in a shared manufacturing arrangement, CBER must be notified and, if these are judged to be important, a PLA and/or ELA amendment must be submitted by all participants and approved by the FDA.

One of the unique requirements in a shared manufacturing arrangement is that each participating manufacturer must perform "significant" manufacturing steps in the production of the licensed product. Until late 1992, CBER had not specifically defined or described the criteria necessary for a manufacturing step to be considered "significant." In its November 1992 policy statement, CBER established that eligibility for separate PLAs and ELAs under shared manufacturing arrangements will be considered when each participating manufacturer performs "manufacturing steps on the product which may alter structure and specificity and that may affect its safety, purity, or potency" (25). Such steps might include, but are not limited to: "(1) Inoculation of vessels or animals for production, (2) cell culture production and characterization, (3) fermentation and harvesting, (4) isolation, (5) purification, and (6) physical and chemical modifications."

A facility's involvement in ancillary manufacturing steps, such as testing, filling, and packaging generally have not been sufficient to warrant eligibility for a shared manufacturing arrangement. In its policy statement on cooperative manufacturing arrangements, however, CBER allows that it may now consider such end-stage activities to meet the eligibility criteria (26). There is one condition: the manufacturer performing these steps must have been responsible for conceiving and developing the product "through extensive preclinical and clinical testing." In other words, CBER now will consider a manufacturer that performs several final manufacturing steps—and that was instrumental in the development of the product—to be eligible for a shared manufacturing arrangement.

Currently, CBER expects the manufacturer performing the final manufacturing steps to provide data demonstrating the product's safety and effectiveness. This manufacturer should also be sufficiently knowledgeable of the overall manufacturing process to be able to identify any manufacturing problems or errors.

CBER also expects the end-product manufacturer to be primarily responsible for post-approval obligations, such as performing stability studies and reporting adverse experiences.

Besides freeing biologics sponsors from the burden of undertaking every manufacturing step, shared manufacturing arrangements can provide other advantages. Participating manufacturers can maintain some degree of confidentiality regarding the manufacturing steps they perform, since the information submitted to CBER to support the restricted PLA need not be shared with other participating manufacturers. However, keeping important manufacturing-related information from participating manufacturers often proves difficult in practice.

A participating manufacturer providing a specialized, partially processed material could conceivably provide this material to additional manufacturers under separate shared manufacturing arrangements. Given their competitive interests, final product manufacturers generally limit the frequency of such situations.

Initially, labeling requirements for shared manufacturing arrangements addressed only partially processed material. The FDA has indicated that certain statements must appear on material destined for further manufacturing (27). The labeling for biological products prepared in a shared manufacturing arrangement must identify all participating licensed manufacturers. However, the package label need only include the name, address, and license number of the end-product manufacturer. The names, addresses, and license numbers of the other participant(s) in the shared manufacturing arrangement need only be included in the description section of the package insert.

Contract Manufacturing

The FDA has always acknowledged that it is common pharmaceutical industry practice to contract for the performance of certain manufacturing operations such as lyophilization, sterilization, and sterile filling (28). CBER also has recognized that there may be certain manufacturing operations that can be "contracted" to another manufacturer without affecting the safety, purity, or potency of a biological product.

Licensed biological manufacturers seeking to employ another manufacturer to perform a part of the manufacturing process for a particular biological product(s) must notify CBER through a PLA or ELA amendment. These amendments must provide sufficient descriptions and details of the contract manufacturing arrangement and must be reviewed and approved in the same manner as any other PLA or ELA amendment.

The manufacturer performing these contract activities need not hold either a PLA or ELA for a biological product. However, if a drug master file or relevant PLA or ELA does exist, CBER generally will request authorization to review such information in support of the contract manufacturing arrangement. An inspection of the contract manufacturing facility or a review of previous CBER or FDA inspections is a possibility. In addition, CBER generally will request a copy of the signed and dated contract entered into by the participating manufacturers and may also require that the agency be notified of the specific production lots subject to the arrangement.

CBER's principal concern in contract manufacturing arrangements is the adequate demonstration of the control and responsibility for the product. Without question, that control and responsibility must remain with the license holder. The license holder can demonstrate such control and responsibility expeditiously through a "man-in-the-plant" type of arrangement, although this has not been a strict requirement for approval. This designated person oversees the contracted operation to ensure that this process is performed as specified in the PLA amendment under Current Good Manufacturing Practices (cGMP) and in the facility described in the ELA amendment. (29)

The existence of a contract manufacturing arrangement is normally known only to the participating parties and CBER. No reference to such an arrangement is required to appear in the labeling for the product(s).

Because many new biological products are developed and manufactured by small pharmaceutical companies, the manufacturing resources available to them are often somewhat limited. For such firms, contract manufacturing arrangements are much easier and more economical than constructing facilities to perform manufacturing functions such as sterilization, lyophilization, and final container filling. Contract manufacturing arrangements may be especially useful in completing the required characterizations of cell lines or microorganisms used in the production process and in the testing of both process intermediates and the final product. Many contract testing laboratories expert in providing such services have emerged to keep pace with the current industry demand.

Divided Manufacturing

Divided manufacturing is the oldest type of joint manufacturing arrangement allowed by CBER. The FDA does not restrict the assignment of production responsibilities in such arrangements, and manufacturers are free to transfer production responsibilities at any step in product manufacture.

There are several unique prerequisites for divided manufacturing arrangements, some of which significantly reduce their appeal to industry. To initiate divided manufacturing, each firm (two or more) must enter into a formal agreement and have the capacity to perform the entire manufacturing process. Also, each firm must have submitted and gained FDA approval for its own PLA and ELA for the subject product and facilities.

The FDA requires all participants in divided manufacturing arrangements to amend their respective PLAs and ELAs to include a description and specific details of the proposed arrangement. Any changes to the proposed arrangement or responsibilities must be submitted to CBER as well.

CBER must approve divided manufacturing arrangements in writing before implementation. In its November 1992 policy statement on cooperative manufacturing arrangements, the center indicated that its staff will assess several factors regarding divided arrangements, including: conformance to commitments specified in relevant PLAs; demonstration of the stability of intermediate products during shipping; the adequacy of labeling for both intermediates and the finished product; and the acceptability of handling clinical and nonclinical complaints concerning the marketed product. Recordkeeping requirements for products made in this manner, and the responsibility for handling and maintaining such records, are described by regulation (30), as are labeling requirements (31).

CBER discussed its interpretation of this labeling regulation in its policy statement. According to that statement, the outer package of products made in divided manufacturing arrangements should include the name, address, and license number of the finished product manufacturer. The description section in package inserts for such products should include the names, addresses, and license numbers of other participating manufacturers. Labeling for all product intermediates should include a statement that the material is intended for further manufacture.

Presently, divided manufacturing arrangements are uncommon, probably because many biological products are produced by a very limited number of competitive manufacturers. Furthermore, there are other alternatives available to manufacturers not wanting or able to perform every step of a complex manufacturing process.

Divided manufacturing arrangements are not extinct. They may, however, be relegated largely to older biological products, such as some vaccines and blood and allergenic products. Several other divided manufacturing arrangements have been or are currently in effect, although these arrangements are not well publicized.

References

(1) 21 CFR Part 600.3(u), April 1, 1992.

(2) Guideline for Drug Master Files, September 1989; Center for Drug Evaluation and Research, Food and Drug Administration.

(3) 21 CFR Part 600.10, April 1, 1992.

(4) 21 CFR Part 600.13 (e)(3), April 1, 1992.

(5) 21 CFR Part 600.13 (e)(2), April 1, 1992.

(6) Guideline on General Principles of Process Validation, May 1987; Center for Drugs and Biologics and Center for Devices and Radiological Health, Food and Drug Administration.

(7) 21 CFR Part 600.10 (b), April 1, 1992.

(8) 21 CFR Part 25.31 (a), April 1, 1992.

(9) Pharmaceutical Manufacturers Association and U.S. Food and Drug Administration Center for Drug Evaluation and Research. Joint Seminar on Environmental Assessments. Rockville, Md., July 29-30, 1991.

(10) 21 CFR Part 601.2 (a), April 1, 1992.

(11) 21 CFR Part 601.12, April 1, 1992.

(12) 21 CFR Part 600.21, April 1, 1992.

(13) Monahan, T.R., Biologics Inspections, 2 Regulatory Affairs. 391. (1990).

(14) 21 CFR Part 601.4 (a), April 1, 1992.

(15) 21 CFR Part 601.5 (b), April 1, 1992.

(16) 21 CFR Part 601.5 (a) , April 1, 1992.

(17) 21 CFR Part 601.6, April 1, 1992.

(18) 21 CFR Part 601.7, April 1, 1992.

(19) 21 CFR Part 601.8, April 1, 1992.

(20) Brady, R.P. and D.A. Kracov, From Diphtheria Antitoxin to Cytokine Products: A Remarkable Scientific Journey/An Aging Regulatory Framework, 3 Regulatory Affairs 105. (1991).

(21) Hill, Regulating Biotechnology Licensed Products, 43 J. Parental Science & Technology 139-41 (1989).

(22) Monahan, T.R., Joint Manufacture of Biological Products in the United States, 2 Regulatory Affairs. 161. (1990).

(23) Federal Register 57,228, FDA's Policy Statement Concerning Cooperative Manufacturing Arrangements for Licensed Biologics (November 25, 1992).

(24) Federal Register 50,795 (November 3, 1983).

(25) Federal Register 57, 228, FDA's Policy Statement Concerning Cooperative Manufacturing Arrangements for Licensed Biologics (November 25, 1992).

(26) Ibid.

(27) Federal Register 50,795 (November 1983).

(28) 21 CFR Part 201.1 (d), April 1, 1992.

(29) Monahan, T.R., Joint Manufacture of Biological Products in the United States, 2 Regulatory Affairs. 161. (1990).

(30) 21 CFR Part 600.12 (e), April 1, 1992.

(31) 21 CFR Part 610.63, April 1, 1992.

DEPARTMENT OF HEALTH AND HUMAN SERVICES	Form Approved; OMB No. 0910-0124
PUBLIC HEALTH SERVICE	Expiration Date: May 31, 1992.
FOOD AND DRUG ADMINISTRATION	See OMB Statement on Page 9
APPLICATION FOR ESTABLISHMENT LICENSE	DATE SUBMITTED
FOR MANUFACTURE OF BIOLOGICAL PRODUCTS	

NOTE: This report is mandated by Section 351 of the Public Health Service Act, the Federal Food, Drug, and Cosmetic Act, Section 502 and Title 21 CFR Part 600. No license may be granted unless this completed application form has been received.

GENERAL INSTRUCTIONS

Type or print legibly in ink. Complete all items. Items which are not applicable, enter 'NA'. If more space is needed for any item continue on an 8 1/2 x 11 inch sheet, reference the entry by item number, and attach. Allow 1 inch top margin for filing purposes. Submit the original and yellow copies of the completed application. Assemble and staple each set, including all attachments. The application forms must be dated and signed by the responsible head. Return the application to DHHS/PHS, FDA/Director, Center for Biologics Evaluation and Research (HFB-240), 8800 Rockville Pike, Bethesda, MD 20892.

I.	**GENERAL INFORMATION**

1a. NAME AND ADDRESS OF MANUFACTURER FOR WHICH U.S. LICENSE IS BEING MADE

CHECK ONE:
☐ NEW APPLICATION
☐ REVISED APPLICATION

b. TELEPHONE NUMBER ⎯⎯⎯ ⎯⎯⎯ ⎯⎯⎯

2. NAME, ADDRESS AND REGISTRATION NUMBER OF EACH LOCATION OF ESTABLISHMENT WHERE ACTUAL MANUFACTURE, INCLUDING TESTING, LABELING AND STORAGE TAKES PLACE. IDENTIFY TYPE OF MANUFACTURING PERFORMED AT EACH LOCATION.

3. NAME AND ADDRESS OF EACH LOCATION FOR COLLECTION OF SOURCE MATERIAL, IF OWNED BY MANUFACTURER (including allergenic product, plasma and blood centers). ALSO INCLUDE LOCATIONS WHERE SOURCE MATERIAL IS OBTAINED UNDER SHORT SUPPLY PROVISIONS (601.22).

4a. NAME AND ADDRESS OF RESPONSIBLE HEAD TO WHOM ALL OFFICIAL CORRESPONDENCE SHOULD BE DIRECTED.

b. TELEPHONE NUMBER ⎯⎯⎯ ⎯⎯⎯ ⎯⎯⎯

5. NAME AND ADDRESS OF OTHER RESPONSIBLE OFFICIAL OR AGENT TO WHOM COPIES OF ALL OFFICIAL CORRESPONDENCE SHOULD BE DIRECTED.

FORM FDA 3210 (5/91)   **PREVIOUS EDITION IS OBSOLETE**   **PAGE 1 OF 9 PAGES**

Biologics Development: A Regulatory Overview

II.	BUILDING AND FACILITIES

1. OUTLINE THE SPECIFIC DESIGN AND CONSTRUCTION FEATURES OF THE VARIOUS MANUFACTURING AREAS INCLUDING TYPE OF PAINT USED, CEILING, WALL AND FLOOR MATERIALS, WHICH FACILITATE CLEANING AND MAINTENANCE.

2. BRIEFLY DESCRIBE THE FUNCTIONS CARRIED OUT IN EACH ROOM WHERE STORAGE, LABELING OR MANUFACTURING, INCLUDING TESTING OCCURS.

3. DESCRIBE THE CONSTRUCTION OF AREAS WHERE STERILE OPERATIONS ARE PERFORMED. WHICH FUNCTIONS ARE CONDUCTED IN THE SAME STERILE AREAS? HOW IS CONTAMINATION WITHIN THESE AREAS KEPT TO A MINIMUM? HOW ARE STERILE AREAS MONITORED FOR CONTAMINANTS?

4. DESCRIBE THE FLOW OF PERSONNEL INTO AND OUT OF EACH AREA. ARE CERTAIN PERSONNEL LIMITED TO CERTAIN AREAS?

II.	BUILDING AND FACILITIES *(Cont'd)*
5. DESCRIBE WATER SUPPLY INCLUDING SPECIFICATIONS FOR EACH AREA OF USE AND SYSTEM VALIDATION.	
6. DESCRIBE AIR SYSTEMS INCLUDING SPECIFICATIONS FOR EACH AREA OF USE AND SYSTEM VALIDATION. WHAT ARE THE TEMPERATURE, HUMIDITY AND AIR PRESSURE DIFFERENTIALS IN EACH AREA?	
7. DESCRIBE METHODS OF CONTROL OF INSECTS AND OTHER PESTS.	
8. DESCRIBE BUILDING MAINTENANCE INCLUDING CLEANING AND REPAIR SCHEDULES.	
9. DESCRIBE USES MADE OF OTHER PARTS OF BUILDINGS OR OTHER BUILDINGS ON THE PREMISES.	
10. DESCRIBE GENERAL SURROUNDINGS OF ESTABLISHMENT AND DISTANCE FROM ANY FARM ANIMALS.	
11. DESCRIBE OTHER INDUSTRY IN VICINITY.	

III.	ANIMAL FACILITIES
1. WHAT ANIMALS ARE USED IN THE PRODUCTION OR TESTING OF PRODUCTS *(Include the types of animals and how they are used for each specific product)*?	

III.	ANIMAL FACILITIES *(Cont'd)*

2. GIVE ADDRESS OF LOCATIONS AND OUTLINE CONSTRUCTION FEATURES OF ALL ANIMAL FACILITIES USED IN MANUFACTURE AND TESTING *(Include a description of water supply, ventilation and capacity of each room).*

3. DESCRIBE ANIMAL QUARANTINE PROCEDURES.

4. DESCRIBE DISPOSAL OF ANIMAL EXCRETA.

5. DO THE SAME PERSONNEL WORK IN ANIMAL AND MANUFACTURING AREAS ON THE SAME DAY?

IV.	WORK WITH MICROORGANISMS

1. WHAT MICROORGANISMS, INCLUDING VIRUSES, ARE BROUGHT INTO OR KEPT IN THE ESTABLISHMENT AND FOR WHAT PURPOSE?

2. WHAT PRECAUTIONS ARE TAKEN WITH THESE MICROORGANISMS?

3. ARE CONTAINERS OF SPORE-BEARING MICROORGANISMS PERMANENTLY MARKED AND KEPT IN A SEPARATE BUILDING DESIGNATED FOR WORK WITH THESE ORGANISMS?

4. DESCRIBE WORK WITH SPORE-BEARING MICROORGANISMS.

V.	EQUIPMENT

1. INCLUDE A PROFILE OF THE MAJOR EQUIPMENT USED IN MANUFACTURE AND TESTING. SPECIFY EQUIPMENT BY MANUFACTURER'S NAME AND LOCATION IN THE WORK AREA.

FORM FDA 3210 (5/91)   **PAGE 4 OF 9 PAGES**

V.	EQUIPMENT *(Cont'd)*

2. WHAT MATERIAL AND EQUIPMENT UNDERGO STERILIZATION BY AUTOCLAVES? *(Give temperature and duration of this temperature in autoclave sterilization, with precautions to insure saturation)*

3. WHAT MATERIALS AND EQUIPMENT UNDERGO DRY HEAT STERILIZATION? *(Give temperature and duration of this temperature in dry heat sterilization, with precautions to insure permeation)*

4. WHAT MATERIALS AND EQUIPMENT UNDERGO STERILIZATION METHODS OTHER THAN THOSE DESCRIBED IN ITEMS 2 AND 3 ABOVE? *(Give conditions of methods and precautions to insure saturation)*

5. DESCRIBE METHODS OF RECORDING TEMPERATURE OF REFRIGERATORS, FREEZERS AND INCUBATORS AND FREQUENCY OF CALIBRATION.

6. DESCRIBE METHODS USED TO CLEAN AND VALIDATE EQUIPMENT USED IN PROCESSING.

FORM FDA 3210 (5/91)

Biologics Development: A Regulatory Overview

V.	EQUIPMENT *(Cont'd)*
7. HOW OFTEN IS EQUIPMENT CLEANED AND CALIBRATED?	

VI.	PRODUCTION AND TESTING
1. GIVE NUMBER OF ROOMS IN WHICH MORE THAN ONE PRODUCT IS PRODUCED *(List products and methods used to prevent cross contamination).*	
2. DESCRIBE FILLING ROOMS AND METHODS USED TO PREVENT CONTAMINATION OF PRODUCTS INCLUDING SANITATION, VALIDATION AND AIR QUALITY DESIGNATIONS.	
3. HOW AND WHERE ARE UNLABELED FINISHED PRODUCTS STORED AND HANDLED?	
4. DESCRIBE METHODS USED TO PREVENT ERRORS DURING THE LABELING OF PRODUCTS.	

FORM FDA 3210 (5/91)

PAGE 6 OF 9 PAGES

VI.	PRODUCTION AND TESTING *(Cont'd)*
5. DESCRIBE PROCEDURE FOR SELECTION AND STORAGE OF RETENTION SAMPLES.	
6. IS DIAGNOSTIC WORK CARRIED ON IN THE LABORATORY? IF SO, UNDER WHAT CONDITIONS?	
7. WHAT OTHER PRODUCTS ARE PRODUCED OR TESTED ON THE PREMISES?	

VII.	RECORDS
1. DESCRIBE MANUFACTURING RECORDS AND ATTACH SAMPLES.	
2. DESCRIBE STERILIZATION RECORDS AND ATTACH SAMPLES.	
3. DESCRIBE DISTRIBUTION RECORDS AND ATTACH SAMPLES.	
4. HOW LONG ARE RECORDS KEPT?	
5. HOW ARE RECORDS MAINTAINED TO PERMIT EFFECTIVE RECALL FROM DISTRIBUTION?	

Biologics Development: A Regulatory Overview

VII.	RECORDS *(Cont'd)*

6. DO RECORDS SHOW THAT EACH STEP IN PROCESSING AND TESTING IS DATED AND SIGNED BY A RESPONSIBLE PERSON?

7. HOW ARE RECORDS INTEGRATED SO THAT, STARTING WITH THE LOT NUMBER, COMPLETE FOLLOW-UP MAY BE MADE OF A PARTICULAR LOT?

8. WHAT IS THE SIGNIFICANCE OF DIFFERENT PORTIONS OF THE LOT NUMBER ON THE FINAL CONTAINER?

9. LIST THE NAMES OF BIOLOGICAL PRODUCTS MANUFACTURED AND APPROXIMATE NUMBER OF LOTS OF EACH COMPLETED DURING THE PRECEDING 12 MONTHS.

10. HAVE ANY LICENSABLE PRODUCTS MANUFACTURED ELSEWHERE BEEN SOLD BY THIS ESTABLISHMENT, OR ANY LICENSABLE PRODUCTS PARTIALLY OR COMPLETELY MANUFACTURED BY THIS ESTABLISHMENT BEEN SOLD TO OTHER ESTABLISHMENTS TO BE COMPLETED, BOTTLED OR LABELED? IF SO, WHEN AND UNDER WHAT CONDITIONS? LIST ANY INTERMEDIATE PRODUCTS *(pastes, powders, etc.)* SHIPPED IN EXPORT.

FORM FDA 3210 (5/91) PAGE 8 OF 9 PAGES

VII.	ORGANIZATION AND PERSONNEL *(Submit the following)*

1. AN ORGANIZATIONAL CHART CONSISTING OF NAMES AND TITLES OF KEY PERSONNEL AND CONSULTANTS.

2. A CURRICULUM VITAE FOR EACH PERSON LISTED IN VII, 1 ABOVE.

VII. ADDITIONAL INFORMATION AND MATERIALS TO BE SUBMITTED FOR THE COMPLETE FILING OF APPLICATION. *(Please state if attached or submitted separately).*

1. Submit floor plans and facility layout which should show:
   a. Step by step flow of raw materials, products and personnel throughout the facility.
   b. Location of major equipment.
   c. Size *(dimensions)* of areas.
   d. Air and water handling systems.
   e. Lavatory and dressing room facilities.

2. Copy of Certificate of Incorporation *(initial submission only)*.

3. Submit an environmental impact analysis report as required by 601.2

IX. COMMENTS

X.	NAMES AND TITLES OF EXPERTS RESPONSIBLE FOR THE PRODUCTION AND TESTING OF PRODUCT	
TITLE	TYPED NAME	SIGNATURE

CERTIFICATION

I certify that all statements made in the application are true and correct to the best of my knowledge and ability. I am familiar with the pertinent sections of Title 21, Code of Federal Regulations, and am aware of my responsibilities described therein.
WARNING: A willful false certification is a criminal offense, U.S. Code, Title 18, Section 1001

SIGNATURE OF RESPONSIBLE HEAD	TYPED NAME AND TITLE	DATE

Public reporting burden for this collection of Information is estimated to average 12 hours per response, including the time for reviewing instructions, searching existing data sources, gathering and maintaining the data needed, and completing and reviewing the collection of information. Send comments regarding this burden estimate or any other aspect of this collection of information, including suggestions for reducing this burden to:

Reports Clearance Officer, PHS
Hubert H. Humphrey Building, Room 721-B
200 Independence Avenue, S.W.
Washington, DC 20201
Attn: PRA

and to:

Office of Management and Budget
Paperwork Reduction Project (0910-0124)
Washington, DC 20503

Please DO NOT RETURN your questionnaire to either of these addresses.

FORM FDA 3210 (5/91)   **PAGE 9 OF 9 PAGES**

Chapter 10
The PLA and ELA Review

by Mark Mathieu
PAREXEL International Corporation

CBER's reorganization and management initiatives affected no other center function as fundamentally as the review and approval process for PLAs and ELAs. The efforts have brought not only significant structural changes, but also a product review process that emphasizes advance planning and application tracking.

The "re-engineering" of CBER's PLA/ELA review process had, in the view of center management, become a necessity in the early 1990s. At that time, CBER management recognized several ominous trends, including steadily climbing average PLA and ELA review times. According to the Pharmaceutical Manufacturers Association (PMA), CBER took an average of 35.3 months to review the six "important" new biological products approved in 1992. This was due largely to the fact that biologics were becoming more complex and, therefore, premarketing submissions were growing in size and complexity

Just as importantly, PLA and ELA submissions were also threatening to become more numerous in the mid- and late 1990s. From 1989 to 1992, the number of biological INDs more than doubled, due largely to a virtual tripling of IND submissions for biotechnology products over the same period. While this increased demands on CBER's IND reviewing machinery, it had a more threatening implication: because these IND submissions represented a future generation of PLAs and ELAs, license applications were likely to surge as the millenium approached. And since PLAs and ELAs put much greater demands on CBER's scientific and reviewing resources than INDs, center management began to prepare for the inevitable.

Such concerns spurred FDA and legislative initiatives that have reshaped CBER's review processes for biological products in several principal ways:

- Under the Prescription Drug User Fee Act of 1992, CBER has, for the first time, specific timelines for PLA and ELA reviews. Starting with submissions made in fiscal year 1994 (October 1993-September 1994), CBER must review and act on 55 percent of complete priority PLAs and ELAs within six months. The target percentage rises yearly through 1997, when CBER must review and act on 90 percent of priority PLAs and ELAs within six months and "standard" license applications within 12 months.

- Also under the Prescription Drug User Fee Act of 1992, CBER will receive funds allowing the center to hire an estimated 300 additional staffers for product-review activities over the next several years.

- A newly implemented refusal-to-file procedure features more specific minimum standards for original PLA and ELA submissions. The procedure is designed to focus CBER reviewing resources on PLAs and ELAs that are materially complete and, therefore, that provide sufficient information on which CBER can base a decision within the newly defined review timelines.

- A new "managed review" process for PLA and ELA reviews will include an early-stage meeting, at which point a division decides whether to accept or refuse to file a new license application, and mid-review licensing committee meetings to assess progress and pending issues. This process also includes an upgraded application-tracking system.

These initiatives do nothing to change the criteria that a sponsor must meet to obtain product and establishment licenses, however. A PLA must show that a biologic is effective, potent, reasonably safe, and sufficiently pure, and that its manufacture is consistent. In addition, the ELA must show that a manufacturing facility—its construction, design, layout, validated processes, environmental monitoring, and other key facility characteristics—employs proper controls and complies with current Good Manufacturing Practice requirements for the biologic it produces. Since CBER will not accept a PLA without its corresponding ELA, the submissions must be made simultaneously.

CBER's New "Managed Review" Process for License Applications

Aside from the obvious structural differences brought about by the reorganization, CBER's new managed review process for ELAs and PLAs is characterized by several features. These include advance planning for and scheduling of reviews, tighter application tracking, and closer communication within and between licensing review committees. The tighter controls are seen as essential if CBER is to meet the ambitious review time frames set by the Prescription Drug User Fee Act of 1992 (see discussion below).

It is important to preface any discussion of CBER's new managed review process by acknowledging that the process is likely to evolve significantly in coming years. And given differences in resources, workload, and the nature of the products regulated, CBER's various offices and divisions are certain to interpret and apply the general concepts of the managed review process somewhat differently. At the same time, the articulation of a defined review process by top CBER officials will, no doubt, produce a more uniform application review process within the center.

Pre-PLA Meetings   Due to the complexity of license applications, CBER makes "pre-PLA meetings" available to sponsors to begin a dialogue intended to promote the submission of high-quality applications. CBER encourages sponsors to request such meetings, particularly companies submitting applications for high-priority products.

Although there is no written guidance on pre-PLA meetings, the FDA's regulations regarding pre-NDA meetings for drugs are applicable in many respects: "[The primary purpose of pre-NDA meetings] is to uncover any unresolved problems, to identify those studies that the sponsor is relying on as adequate and well-controlled to establish the drug's effectiveness, to acquaint FDA reviewers with the general information to be submitted in the marketing application (including technical information), to discuss appropriate methods for statistical analysis of the data, and to discuss the best approach to the presentation and formatting of data in the marketing application."

Timing is one of the factors that can influence the utility of a pre-PLA meeting. To promote useful discussions, CBER officials recommend that pre-PLA meetings be held after the sponsor has collected most or all of the data and information pertinent to a biologic's safety and effectiveness.

CBER officials also suggest that sponsors, when requesting a meeting, be as specific as possible regarding the issues to be discussed, the attendees deemed necessary for the discussion, and the sponsor's goals for the meeting (e.g., to reach agreement on the presentation of clinical data).

Either the pre-PLA meeting or the submission of the PLA and ELA will trigger several "auxiliary actions" that, in effect, represent the beginning of the managed review process. In either case, CBER will begin to evaluate the need for, and forecast the scheduling of, key activities that will become critical later in the review process. First, CBER will assess whether a bioresearch monitoring inspection might be necessary to verify data and information included in the license applications. Center staff will also review the meeting schedule of the FDA advisory committee relevant to the application to forecast possible dates for committee presentations. While CBER officials concede that such early efforts represent a challenge, they point out that this advance planning will ultimately result in more efficient reviews.

Like the IND review, the license review process begins in CBER's Document Control Center. There, the applications are "triaged"—staffers log key information about the PLA and ELA (e.g., sponsor name, product, receipt date) and determine the offices to which the applications should be forwarded.

While the reorganization has consolidated IND and PLA reviews into product-specific research and review offices, it has separated ELA reviews from product evaluations. The ELA is forwarded to the Division of Establishment Licensing within CBER's Office of Establishment Licensing and Product Surveillance, while the PLA is forwarded to one of three divisions:

- the Division of Vaccine and Related Products Applications (DVRPA) within the Office of Vaccine Research and Review;

- the Division of Application Review and Policy (DARP) within the Office of Therapeutics Research and Review; or

- the Division of Blood Establishment and Product Applications (BEPA) within the Office of Blood Research and Review.

CENTER FOR BIOLOGICS EVALUATION AND RESEARCH

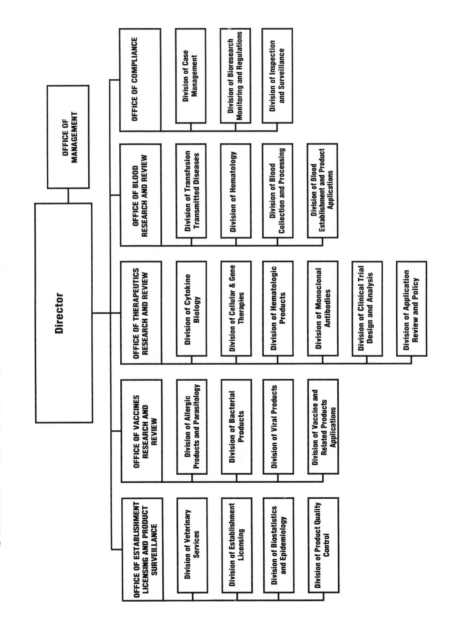

Initial PLA and ELA Processing   With few exceptions, initial processing of PLAs and ELAs is quite similar.

Once the PLA is within BEPA, DARP, or DVRPA, a CSO or scientist will perform various administrative processing and screening activities. For example, a staffer screens the PLA to determine if any major deficiencies exist and attempts to verify that the applicant has paid the required fees associated with the submission. Applications meeting basic submission requirements are assigned a reference number for tracking purposes. Also, the division prepares an acknowledgement letter informing the sponsor of the application's reference number, agency contact person, and other information.

During this initial processing phase, several other important activities take place. If not done at the pre-PLA stage, the division will attempt to designate the PLA as either a priority (i.e., a clear advance relative to existing products) or standard (i.e., all others) product application. Also, the division determines whether the application will require a collaborative review with any other centers within the FDA (e.g., the Center for Drug Evaluation and Research).

In most respects, the Division of Establishment Licensing's initial processing of the ELA mirrors early PLA processing. The ELA is subjected to an administrative processing and screening process, after which it receives its own reference number.

However, the Division of Establishment Licensing is not involved in decisions regarding either product designation or intercenter collaboration. The office handling the PLA will notify the division whether the sponsor has paid the necessary application fee.

Licensing Committees   CBER's reorganization did not affect the manner in which PLAs and ELAs will be reviewed. Product reviews will continue to be conducted by two licensing committees: the PLA licensing committee and the ELA licensing committee.

Following the initial processing of the license applications, the PLA is forwarded to the chairperson of the PLA licensing committee and the ELA is forwarded to the chairperson of the ELA licensing committee. The process of identifying and assigning chairpersons, who are generally scientists or physicians, is likely to have started during the pre-PLA phase. Such assignments are typically based on several factors, including expertise and availability. The PLA chairperson will most likely have been involved in the review of the IND for the product.

For the ELA, the Division of Establishment Licensing assigns a chairperson from one of its three branches: Branch I, which handles applications for bacterials, cytokines, and test kits; Branch II, which reviews ELAs for hematology products, therapeutics, and fractionation products; or Branch III, which reviews applications for viral vaccines and monoclonal antibodies.

The PLA's path is slightly less straightforward. Depending on the product's nature, the initial processing of the PLA will be conducted by either DVRPA, DARP, or BEPA. Although each of these divisions is staffed, in part, by scientists, the PLA chairperson will, in most cases, work in one of the research and review divisions in CBER's three product-specific offices:

- within the Office of Vaccine Research and Review, DVRPA forwards the PLA to a chairperson within either the Division of Allergic Products and Parasitology, the Division of Bacterial Products, or the Division of Viral Products (for a more detailed analysis of the products these divisions regulate, see Chapter 5);

- within the Office of Therapeutics Research and Review, DARP forwards the PLA to a chairperson within either the Division of Cytokine Biology, the Division of Cellular and Gene Therapies, the Division of Hematologic Products, or the Division of Monoclonal Antibodies (for a more detailed analysis of the products these divisions regulate, see Chapter 5);

- within the Office of Blood Research and Review, BEPA forwards the PLA to a chairperson within either the Division of Transfusion Transmitted Diseases, the Division of Hematology, or the Division of Blood Collection and Processing (for a more detailed analysis of the products these divisions regulate, see Chapter 5).

Generally, the ELA and PLA chairpersons first scan their respective applications to determine the nature and complexity of the product and the submission. This screening is particularly important because it allows the chairpersons to identify the types of expertise needed in assembling the PLA licensing committee and the ELA licensing committee. The individuals selected for the respective committees comprise the team of reviewers who will evaluate the applications.

The size and composition of PLA and ELA licensing committees can vary significantly, influenced by such factors as the nature of the product under review and the scientific and technical issues that its review is likely to present. The PLA licensing committee, for example, will consist of the chairperson, at least one clinical reviewer, a product manufacturing reviewer, a statistician (from CBER's Division of Biostatistics and Epidemiology), and a pharmacologist/toxicologist. Generally, the scientist who served as the lead reviewer for the product's IND will also sit on the committee.

ELA licensing committees are typically comprised of biologists, microbiologists, and pharmacists. A member of the PLA licensing committee also serves on the ELA committee.

Licensing committee membership will also differ somewhat across CBER's various offices. Licensing committees within the Office of Therapeutics Research and Review, for example, will usually include a physician from that office's Division of Clinical Trial Design and Analysis to analyze clinical trial data and designs.

When outside expertise is necessary, the licensing committee may include members from other FDA centers. The FDA has published intercenter agreements for three centers—CBER, the Center for Drug Evaluation and Research, and the Center for Devices and Radiological Health—whose product review authorities sometimes overlap (see Chapter 14). These documents do the following: (1) they assign to the centers jurisdiction for the regulation of specific biological, drug, device, and combination products; (2) they identify those product characteristics and indications that require collaborative review efforts; and (3) they discuss logistical aspects of collaborative reviews.

Either separately or jointly, the two licensing committees then conduct their first meeting. At these meetings, the committee chairpersons formally assign responsibilities to committee members. Also, each committee identifies intercenter consultations (if any are needed), and identifies and discusses major regulatory issues (e.g., clinical, policy, product manufacturing) regarding the product review. The PLA licensing committee is responsible for verifying that auxiliary actions—regarding bioresearch monitoring inspections and possible advisory committee meetings—have been undertaken.

Committee members must then assess the applications to determine if they should be "filed"—that is, whether the applications meet basic standards of completeness and quality to justify a formal review. PLAs or ELAs not meeting these standards will become subject to a "refusal-to-file" (RTF) decision, which means that CBER will discontinue the review process. By refusing deficient submissions and eliminating the time-consuming review efforts associated with them, CBER hopes to focus its strained resources on the highest quality PLAs and ELAs.

In July 1993, CBER released its first RTF guidance document to identify the criteria the center uses to determine if ELAs and PLAs are acceptable. The guidance is consistent with CDER's refusal-to-file criteria for NDAs.

Based on the reviews of their members, the ELA and PLA committees must each make a recommendation on the filing of the applications. Under the managed review process, these reviews and decisions must be reached quickly, since committees will meet within either 45 or 60 days of the applications' submission to reach a joint filing decision.

Such meetings are called 45-Day or 60-Day Filing Meetings. For priority biologics, the licensing committees must meet within 45 days of the submission date. For standard applications, the committees must meet within 60 days of the submission.

After informing upper management of the decision reached at the filing meeting, the committee develops a letter informing the sponsor that the applications either have been accepted for filing or refused. If a committee decides to issue an RTF decision, it will document the applications' principal deficiencies.

Assuming the applications are accepted for review, the filing meeting serves as a planning session, during which the reviews are planned, scheduled, and assigned. At that point, the two committees separate for the formal application reviews.

Not surprisingly, licensing committees often function differently. For example, some PLA licensing committees prefer to conduct work sessions in whole-group meetings, while others tend to separate into clinical and manufacturing subgroups.

As the reviews progress, there is generally considerable dialogue between CBER and the sponsor. Questions and concerns raised during the review, for example, will be documented by the committee chairperson and may be formalized in an information request letter sent to the responsible head identified in the application. If a licensing committee expresses serious mid-review concerns about the quality or quantity of information or data submitted in the application, a sponsor-review committee meeting may be scheduled.

Biologics Development: A Regulatory Overview

In part due to the new deadlines imposed for PLA and ELA reviews under user fee legislation (see discussion below), CBER has developed new methods for tracking the status of application reviews and associated time frames. CBER is also tracking product reviews more closely through "mid-point" meetings, when licensing committee members report to senior office staff on the progress of, and pending issues regarding, ELA and PLA reviews.

During the review process, CBER also tests samples from production lots, samples that the sponsor must submit with lot analyses and manufacturing protocols. The samples are tested to assure that the product meets the proposed specifications for potency, purity, and consistency.

The Pre-License Inspection Once the licensing committees' principal questions and concerns have been addressed, CBER performs a pre-license inspection of the manufacturing facility specified in the ELA. A requirement for any new product licensure, the pre-license inspection is normally conducted when the establishment is in active operation and is manufacturing the product for which the ELA and PLA have been submitted. For this reason, the inspection is normally an announced inspection—that is, CBER works with a manufacturer to identify a point at which the facility is in active production and responsible personnel are available so that a meaningful inspection can be made.

Generally, the pre-license inspection team consists of the chairperson of the PLA committee and the chairperson or a member of the ELA committee. An FDA field inspector may accompany these individuals during the inspection.

To verify information submitted in the ELA and PLA, the pre-license inspection is designed as an in-depth investigation and review of the manufacturing operations performed in various departments or processing areas. Focal points of the inspection include the manufacturer's production process and test procedures, and an examination of the facility, equipment, and recordkeeping for compliance with cGMP. The product must also be available for inspection at every stage of manufacture, including the finished product intended for sale.

If significant violations are discovered during the pre-license inspection, the ELA and PLA cannot be approved. An FDA re-inspection will be ordered only after CBER has obtained assurance that the condition(s) cited during the original inspection have been corrected.

Final Stages of the Review Process   Each member of the ELA and PLA licensing committees develops a review memorandum on the relevant aspect(s) of the applications. After reviewing the memoranda, the committee chairpersons consult with the various reviewers to reconcile the evaluations and move toward a decision on the applications.

When the PLA is considered ready for approval, the applications are forwarded to either DARP, BEPA, or DVRPA, which performs a "quality control" check to ensure that the decision is consistent with CBER's powers under existing laws and regulations and with existing center policy. From there, the PLA recommendation is forwarded to the director of the office—the Office of Therapeutics Research and Review, the Office of Blood Research and Review, or the Office of Vaccines Research and Review—regulating the product. ELAs are forwarded to the Office of Establishment Licensing and Product Surveillance. The office directors hold final sign-off authority for new product and establishment licenses.

Communicating the Review Results to the Sponsor   The Prescription Drug User Fee Act of 1992 requires CBER to use "action letters" to communicate the results of PLA and ELA reviews to sponsors. Although the use of so-called action letters is standard practice in CDER's drug review process, it clearly represents a change for CBER.

The user-fee legislation calls for CBER to "act on" PLAs and ELAs within specific time periods—within six months on 55 percent of priority products submitted during fiscal year 1994, for instance. CBER fulfills this requirement by issuing an action letter to the sponsor. The center was expected to institute a formal policy regarding action letters in 1993.

In most important respects, CBER's use of action letters is similar to CDER's use of the three basic types of action letters:

Approvable Letter. CBER forwards an approvable letter "if the application substantially meets the requirements for marketing approval and the agency believes that it can approve the application if specific additional information or material is submitted or specific conditions are agreed to by the applicant." Often, CBER will request in the approvable letter that the sponsor submit safety update reports or final printed labeling (FPL) for the product.

Approval Letter. An approval letter indicates that the subject biologic is considered approved as of the date of the letter. In some approval letters, however, CBER does request minor editorial changes to the product's labeling.

Not-Approvable Letter. A not-approvable letter informs the sponsor that the information in its application is insufficient to justify approval at that time. The letter will identify the deficiencies in the applications and the remaining issues relevant to the product's approval.

PLA and ELA approval—or the licensure of a product and a facility to manufacture that product—is granted in the form of two licenses: a product license and an establishment license. The licenses are issued to the establishment or legal entity engaged in the product's manufacture.

Sponsor Rights During the PLA and ELA Review Process

Biologic sponsors have a variety of rights during the review of their PLAs and ELAs. Two of the more important ones are the right to confidentiality and a timely review.

The Right to Confidentiality Given the quantity of competitively sensitive data and information contained in most PLAs and ELAs, confidentiality is of great importance to product sponsors. At both the federal government and agency levels, there are standards and procedures designed to protect certain types of information from public disclosure.

In general terms, the FDA is prohibited from releasing any information or test data that qualify as either "trade secret" or "commercial or financial information." A trade secret is defined as "any formula, pattern, device, or compilation of information which is used in one's business and which gives him an opportunity to obtain an advantage over competitors who do not know or use it." Commercial or financial information is "valuable data or information which is used in one's business and is of a type customarily held in strict confidence or regarded as privileged and not disclosed to any member of the public" by the company to which it belongs. Included in these definitions are manufacturing processes, chemistry information, and marketing assessments.

For the purposes of identifying information to be protected from disclosure, CBER regulations establish what is called a "biological product file." According to FDA regulations, this file "includes all data and information submitted with or incorporated by reference in any application for an establishment or product license, INDs incorporated into any such application, master files, and other related submissions." At no time before a product license is issued to an applicant may CBER disclose the existence of the biological product file or data or information in the file.

If the file's existence has been publicly disclosed or acknowledged by the sponsor or another source before a license is issued, CBER is still prohibited from disclosing data or information from the file. In this case, however, "the Commissioner may, in his discretion, disclose a summary of such selected portions of the safety and effectiveness data as are appropriate for public consideration of a specific pending issue, e.g., at an open session of a Food and Drug Administration advisory committee or pursuant to an exchange of important regulatory information with a foreign government."

Even after a license is issued, the FDA is still prohibited from disclosing certain types of information in the biological product file, unless: (1) they have been previously disclosed to the public; or (2) they relate to a product or ingredient that has been abandoned and, therefore, no longer represent a trade secret or confidential commercial or financial information. CBER regulations identify these forms of information as: (1) manufacturing methods or processes, including quality control procedures; (2) production, sales, distribution, and similar data and information, except any compilation of such data and information aggregated and prepared in a way that does not reveal data or information which is not available for public disclosure; and (3) quantitative or semiquantitative formulas.

CBER is permitted to release several forms of data and information following the issuance of a license, including:

- all safety and effectiveness data, meaning all studies and tests of a biological product on animals and humans, and all studies and tests on the biologic for identity, stability, purity, potency, and bioavailability;

- a protocol for a test or study, unless it is shown to fall within the exemption established for trade secrets and confidential commercial or financial information;

- adverse reaction reports, product experience reports, consumer complaints, and other similar data and information after the deletion of the names and related information that would identify the person using the biologic or any third party involved with the report, such as a physician or hospital or other institution;

- a list of all active ingredients and any inactive ingredients previously disclosed to the public;

- an assay method or other analytical method, unless it serves no regulatory or compliance purpose and it is shown to fall under the definition of trade secret or commercial or financial information;

- all correspondence and written summaries of oral discussions relating to the biological product file;

- all records showing the manufacturer's testing of a particular lot after deletion of data or information that would show the volume of the biologic produced, manufacturing procedures and controls, yield from raw materials, costs, or other information falling under the definition of trade secret or commercial or financial information; and

- all records showing the FDA's testing of and action on a particular production lot.

The Right to a Timely Review   Unlike CDER, which faced a statutory requirement to complete NDA reviews within 180 days, CBER did not, until recently, have time limits imposed on its reviews of PLAs and ELAs. Within the Prescription Drug User Fee Act of 1992, however, the FDA did make a series of commitments to reduce NDA and PLA/ELA review times in exchange for the additional funding the legislation was to provide the agency. These commitments were made in two letters written by FDA Commissioner David Kessler, M.D., and were incorporated into the law.

Dr. Kessler's letters established a series of interim goals starting in fiscal year (FY) 1994 (i.e., October 1993-September 1994) that lead to final five-year goals to be reached in FY-1997. Beginning with NDAs, PLAs, and ELAs submitted in FY-1997, the FDA's five-year goals were defined as follows:

- to act on 90 percent of complete priority applications and priority amendments within six months of submission (*Editor's note: priority applications are products that appear "to represent a therapeutic advance"*);

- to act on 90 percent of complete standard applications and supplements with clinical data within 12 months of submission (*Editor's note: standard applications are nonpriority applications*);

- to act on 90 percent of complete applications resubmitted after the receipt of a not-approvable letter within six months of resubmission; and

- to act on 90 percent of supplements not requiring the review of clinical data (e.g., manufacturing supplements) within six months of submission.

In the interim, the legislation calls for CDER and CBER to begin fulfilling these review time frames for most applications (i.e., 55 percent), beginning with applications submitted in FY-1994. This percentage will escalate to 70 percent for applications submitted in FY-1995 and to 80 percent of applications submitted in FY-1996. CBER's interim targets also include a goal to eliminate its backlog of overdue PLAs, ELAs, and PLA amendments within 24 months of October 1, 1992, the effective date of user fee payments.

The review time frame for a particular PLA or ELA submission may be extended if the sponsor submits amendments during the review process. Although no extension is mandated for minor amendments, the submission of major amendments, such as a significant amount of new clinical data, will trigger a three-month extension to the target timeline for the review if submitted within three months of the action due date. If the major amendment is a response to a not-approvable letter, however, a six-month extension is necessitated.

Along with these goals, CBER has also established several other interim goals designed to help the center meet the aggressive review deadlines:

- hiring 50 percent of the additional staff provided for by the user fee legislation by the first quarter of FY-1995, and hiring 100 percent of this incremental staff by the end of FY-1997;

- establishing an FDA/industry working group to develop and to oversee joint programs and improve review times;

- implementing project management systems within 18 months of the initiation of user fee payments for applicable PLA and ELA reviews;

- implementing performance tracking and monthly monitoring of CBER performance within six months of the initiation of user fee payments; and

- initiating a pilot computer-assisted product license application review (CAPLAR) program during FY-1993.

Chapter 11:

The FDA's Biological Products Advisory Committees

by Jack Gertzog
Director, Division of Scientific Advisors
Center for Biologics Evaluation and Research
Food and Drug Administration

In reaching regulatory decisions regarding new biological products, CBER's director and professional staff may seek assistance from any one of four scientific advisory committees. When called upon, the physicians, scientists, biostatisticians, and other members of these committees weigh available evidence and provide scientific and medical advice on the safety, effectiveness and appropriate use of products under CBER's jurisdiction.

For several years, CBER's advisory committee process had been coordinated by the Center for Drug Evaluation and Research's (CDER) Office of Advisors and Consultants, which manages CDER's 17 pharmaceutical advisory committees as well. In late 1991, that office's responsibilities were divided, and a Division of Scientific Advisors was established within CBER.

Among this office's goals is to improve the overall quality and responsiveness of the center's biological review process through the use of extramural scientific advisors and consultants in an advisory capacity and to optimize the effective use of the advisory program.

For some years, CBER averaged more meetings per committee than did other components of the FDA. During the past few years, however, reorgani-

zation and senior-level vacancies within CBER resulted in a decline in the
meeting frequency of some biological committees (see table below).
Committee activity has started to increase as these issues are addressed.

In discussing the role of biological advisory committees in the IND, PLA,
and ELA review processes, it is helpful to review the following aspects of the
advisory committee process itself: purpose; scope; meeting process; committee membership and dynamics; use of evidence; findings and recommendations; and dissemination of findings.

CBER Advisory Committee Meeting Frequency

	1987*	1988*	1989*	1990*	1991*	1992*
Allergenic Products Advisory Committee	3	1	0	0	0	1
Biological Response Modifiers Advisory Committee	–	–	–	1	2	3
Blood Products Advisory Committee	3	2	4	4	3	4
Vaccines and Related Biological Products Advisory Committee	4	4	5	3	2	3
Total Meetings	**10**	**7**	**9**	**8**	**7**	**11**

* Fiscal Year

Source: FDA

Purpose

The charter of each advisory committee specifies the purpose of that committee (see exhibit below). For example, the Vaccines and Related Biological Products Advisory Committee (VARBPAC) "reviews and evaluates data relating to the safety, effectiveness and appropriate use of vaccines and related biological products which are intended for use in the prevention, treatment or diagnosis of human diseases, and as required, any other product for which the FDA has regulatory responsibility. . . and makes appropriate recommendations to the Commissioner of Food and Drugs." Each of CBER's other three advisory committees is similarly structured by biologic category—allergenic products, biological response modifiers, blood products—for its specific functions.

Biological advisory committees are also responsible for peer review of the center's intramural research programs, which provide scientific support for biological product regulation. Within the FDA, this peer review function is unique to biological committees. CBER believes that its intramural research program is essential to a high quality, expeditious biological product review process and that its own scientific program must therefore be held to the most exacting peer standards.

Committee Function Statements

Allergenic Products Advisory Committee

The Committee reviews and evaluates available data concerning the safety, effectiveness, and adequacy of labeling of marketed and investigational allergenic biological products or materials administered to humans for the diagnosis, prevention, or treatment of allergies and allergic disease.The Committee makes appropriate recommendations to the Commissioner of Food and Drugs of its findings regarding: the affirmation or revocation of biological product licenses; on the safety, effectiveness, and labeling of the products; clinical and laboratory studies of such products; amendments or revisions to regulations governing the manufacture, testing, and licensing of allergenic biological products; and the quality and relevance of FDA's research programs, which provide the scientific support for regulating these agents.

Biological Response Modifiers Advisory Committee

The Committee reviews and evaluates data relating to the safety, effectiveness, and appropriate use of biological response modifiers intended for use in the prevention and treatment of a broad spectrum of human diseases. The Committee also considers the quality and relevance of the FDA's research program, which provides scientific support for the regulation of these products and makes appropriate recommendations to the Secretary, the Assistant Secretary for Health, and the Commissioner of Food and Drugs.

Blood Products Advisory Committee

The Committee reviews and evaluates available data concerning the safety, effectiveness, and appropriate use of blood products derived from blood and serum or biotechnology intended for use in the diagnosis, prevention, or treatment of human diseases, and, as required, any other product for which the FDA has regulatory responsibility. The Committee advises the Commissioner of Food and Drugs of its findings regarding: the safety, effectiveness, and labeling of the products; clinical and laboratory studies involving such products; the affirmation or revocation of biological product licenses; and the quality and relevance of the FDA's research program, which provides the scientific support for regulating these agents. The Committee will function at times as a medical device panel under the Federal Food, Drug, and Cosmetic Act Medical Device Amendments of 1976. As such, the Committee recommends classification of devices subject to its review into regulatory categories; recommends the assignment of a priority for the application of regulatory requirements for devices classified in the standards or premarket approval category; advises on formulation of product development protocols and reviews pre-market approval applications for those devices to recommend changes in classification as appropriate; recommends exemption of certain devices from the application of portions of the Act; advises on the necessity to ban a device; and responds to requests from the Agency to review and make recommendations on specific issues or problems concerning the safety and effectiveness of devices.

Vaccines and Related Biological Products Advisory Committee

The Committee reviews and evaluates data relating to the safety, effectiveness, and appropriate use of vaccines and related biological products intended for use in the prevention, treatment, or diagnosis of human diseases, and, as required, any other product for which the FDA has regulatory responsibility. The Committee also considers the quality and relevance of the FDA's research program, which provides scientific support for the regulation of these products and makes appropriate recommendations to the Commissioner of Food and Drugs.

On a more general level, the committees fulfill the following roles for CBER:

- they provide necessary and frequently scarce expertise from physicians and scientists not otherwise available to the agency;

- they provide these services on a cost-effective basis (i.e., only as needed);

- they provide the agency with a stronger scientific basis for decisions and actions; and

- they provide the means for, and assure that agency actions are carefully reviewed in, a public forum with a broad spectrum of interested, qualified non-agency participants.

Scope

As stated, CBER has access to four biological product advisory committees. The existence of these committees is not statutorily mandated. CBER decides not only whether a certain committee is needed, but the committee's structure and function as well. With the strong advances in biotechnology and increasing numbers of biotechnology INDs and PLAs being submitted, CBER is evaluating the need for additional committees.

Routine issues are not considered by the committees. Innovation, difficulty or complexity, controversy, breadth of applicability, need for specialized expertise not otherwise available, and health care implications are the general criteria that determine the topics brought to committee (see exhibit below for frequently used criteria to identify PLAs and INDs to be presented to biological advisory committees).

General Criteria for Identifying INDs and PLAs
To Be Presented To Advisory Committees

INDs To Be Presented to Advisory Committees

1. Products that represent important diagnostic, therapeutic, or preventive advances (safety or effectiveness).
2. Products with novel and improved delivery methods.
3. Products with potential or apparently significant safety problems.
4. Products about which significant disagreement exists between the FDA and the sponsor.
5. Products involving new biotechnology.
6. Products in which the committee or public has a significant interest.

PLAs To Be Presented to Advisory Committees

1. Significant new products.
2. Marketed products proposed for significant new uses.
3. Products with significant potential for risk relative to their therapeutic benefit.
4. Products whose effectiveness is under debate by the scientific and medical community.
5. Products under consideration for postmarketing studies.
6. Products that may be withdrawn because of questionable safety or effectiveness.

The issues brought before a committee may be very specific (e.g., evaluate a PLA or reports of increased adverse reactions with the use of a specific marketed biologic) or more general in nature (e.g., development or review of a points-to-consider document or a general discussion on the use of surrogate markers for clinical endpoints). Because CBER knows the current status of the issue and because committee use is optional, the center decides if and when the issue is ready for scientific committee review. However, when a committee expresses a strong interest in a matter, CBER always tries to accede to the committee's wishes and has, to this point, never denied a committee request to meet on a subject.

The agency also carefully considers requests by regulated industry and consumer organizations that particular issues be brought to committee. The FDA has decided both ways on such requests, depending on the facts of the case. The major determining factor in such decisions—given that the committees are primarily scientific and technical groups—is whether there are sufficient scientific data and information concerning an issue.

Meeting Format and Process

Committee meeting dates are frequently scheduled before CBER develops an agenda, and are selected to maximize committee member participation. Within a particular time period identified by the agency (e.g., the last two weeks of June), the members are polled for specific meeting dates. Thus, in virtually all instances, the members select the specific meeting date; the exceptions to this are emergency, short lead-time meetings.

Each committee convenes from one to five times per year—meeting frequency varies from committee to committee and from year to year. Subcommittee meetings also may be held and must comply with the same requirements for full meetings with regard to public notices of meetings, open sessions, and the transcription of proceedings. Federal regulations state that the "subgroup will not be established as a committee distinct from the parent committee. However, a subgroup will be established as a separate committee when the charter of the parent committee does not incorporate the activities of the subgroup." CBER has not scheduled any subcommittee meetings over the past few years.

After CBER identifies specific issues, the committee's designated federal official (DFO)—also called the executive secretary—works with the committee chairperson and the appropriate scientific staff within CBER to develop a detailed agenda. The sponsor and principal investigators of the product to be reviewed are contacted and invited to attend and make a presentation at the meeting. Other interested persons or organizations that might want to participate are also directly notified of the meeting.

CBER publishes meeting notices in the *Federal Register*, the FDA commissioner's calendar, and public information releases. Trade newsletters and some journals routinely publish meeting dates as well. At the time of this writing, the FDA was considering an "800" number to provide information on schedules and agendas.

Meetings are open to the public, except when confidential or trade secret information is reviewed. In all cases, the meeting provides time for the public to present statements on matters pending before a committee. In this context, the "public" might include private citizens and representatives of regulated industry, consumer organizations, federal and state health organizations, and medical, scientific and other professional organizations.

Facilities usually are large enough to accommodate 100 to 300 attendees, and attempts are made to reserve facilities that can accommodate the estimated attendance (up to 400). But it is not mandatory for the FDA to provide space for all interested persons, and factors such as limited funding and availability of space may, on occasion, limit attendance.

Briefing material for a committee is prepared by both the agency and the sponsor of the product. Other involved persons or groups may also be asked to prepare briefing information.

While CBER attempts to make this information available to the committee two to four weeks in advance of the meeting, the center sometimes fails to meet this schedule. In addition to the scientific data and studies, the agency will usually develop an issue/question statement intended to help a committee focus on questions of primary concern to the FDA. The chairperson may expand discussion beyond these "guidelines." Committee members may, of course, raise other questions they consider appropriate.

A committee may designate primary and secondary reviewers on the committee. Such designations are optional and vary with committee and issue.

Generally, the committee does not prepare or submit written reports on specific products. Unedited verbatim transcripts of each meeting, and summary minutes drafted by the DFO and approved by the committee chairperson or full committee, constitute the official record of the meeting. Both documents are available to the public (except for trade secret, proprietary, or other protected information).

Member Selection and Group Dynamics

Candidates for committee vacancies are identified in several ways. The DFOs perform literature searches, attend relevant scientific meetings and symposia, discuss potential candidates with former and current committee members, and contact chairpersons of those medical school departments that are highly ranked or have recognized strengths in needed areas of expertise. Colleagues within the FDA and NIH also are solicited for recommendations. Appropriate medical and scientific societies are contacted for suggestions as well.

Annual committee vacancy announcements are published in the *Federal Register* and are frequently republished in organizational newsletters and journals. Unsolicited inquiries about committee vacancies also are received.

The charter for each committee identifies the structure of the committee and the areas of expertise or specialization that are necessary. The agency may further specify its needs as vacancies occur. This determination is based on the agency's assessment of the specific issues it believes will come before the committee during the next three to four years and a determination of what specialties are needed to achieve a balance of expertise on the committee.

For example, the VARBPAC charter may specify that pediatricians are to be appointed to the committee, but the agency may believe that viral vaccine issues will predominate during the next two to four years. Accordingly, the agency would recruit pediatricians with expertise in pediatric viral infections or neurologists who specialize in viral infection complications. Conversely, if the VARBPAC membership is unbalanced (e.g., perhaps the committee has six pediatricians and no molecular biologist or biostatistician), CBER may decide to fill the vacancies with the underrepresented or unrepresented specialties.

Recently, the FDA modified the charters of its scientific advisory committees to allow supplementation of the voting membership when necessary expertise is not available from existing committee members. These additional voting members must be appointed as special government employees (SGE), and must meet all of the requirements of other SGE committee members. As of this writing, CBER has not used this procedure. When committees have needed additional expertise in the past, CBER has appointed SGE consultants or experts who participated fully in the committee meetings but did not have official voting status.

Other parameters also are considered in selecting committee members. While balance is sought in terms of gender, race, and geographic location, technical competence is the overriding consideration. Although most committee members are physicians, several members have doctorates in the biomedical sciences or both medical and scientific degrees. CBER is in the process of appointing biostatisticians to each of its committees. Two of the CBER committees have members who are consumer representatives.

The "ideal" committee member has both clinical and research experience, and is both health care- and science-oriented. Because terms of appointment are four years in duration and because there are more types of issues than committee members, the "ideal" committee member is one who has both recognized accomplishments and leadership within his or her specialty and a demonstrated ability and interest in issues outside that specialty.

Committee members are appointed as SGEs, and must adhere to federal and agency policy concerning the protection of privileged information, impartiality, objectivity, and conflicts of interest. Each receives a stipend of $150 per day and reimbursement for travel, food, lodging and other costs incidental to meetings.

As stated, terms of appointment generally are four years, except initial appointments to newly established committees, which vary to limit committee turnover to only 25 to 35 percent per year. Terms may be extended beyond four years until a new individual is appointed and functioning as an active committee member.

Committee meetings are led by the chairperson, who the FDA commissioner selects from the committee's members. The chairperson has authority to conduct the meeting, including adjournment, discontinuation of discussion on any particular matter, and any other action deemed necessary for a fair and expeditious meeting. The chairperson is a regular voting member of the committee.

CBER appoints a DFO (executive secretary) for each committee. The DFO is an FDA employee, but is not a member of the committee. As noted, the DFO serves as the executive secretary for the committee, and is the official CBER liaison to the committee. The executive secretary is responsible for all administrative planning and preparation for a meeting, as well as the ancillary steps for establishing and staffing a committee. This individual "must be knowledgeable of FDA's legislative and regulatory mandates and be capable of maintaining a proper scientific or technical direction in order to achieve the most satisfactory results." All communications with the committee should be directed through the executive secretary. A successful committee process depends on an effective working relationship between the chairperson and the executive secretary.

To preserve the independence and objectivity of advisory committees, the FDA has extremely stringent conflict-of-interest (COI) criteria. For this reason, it is virtually impossible to select committee members from regulated industry, although there is no regulatory prohibition against this. Agency policy excludes FDA staff from committee membership.

In general, scientific committee members work in academia. Because many academicians work closely with product sponsors, COI concerns can be a severely limiting factor in recruiting technically qualified committee members in certain areas of expertise.

Use of Information in Product Evaluations

For each biological product evaluated by a CBER committee, relevant data and background information are provided by the sponsor of the product, CBER, and other interested persons or organizations. Written information is submitted to the committee in advance, and presentations are made at the meeting.

If the issue pertains to a PLA, information from the product sponsor should be derived from adequate and well-controlled studies. It is unlikely that a PLA issue will be brought to committee until such data are available or such studies are under way. If the issue concerns a product in an early stage of development or investigation, the committee may become involved in reviewing and evaluating protocol development and study design as well as Phase 1 through Phase 3 clinical data.

Information presented to a committee may be derived from published or unpublished data. If an IND and its data are as yet unpublished, they may be considered proprietary and confidential, and accordingly, only reviewed or discussed in closed meetings. Premarketing submissions and raw data (i.e., case reports, individual animal data, etc.) are usually voluminous and generally are not provided to the committee. If committee members request raw data, however, it is provided. The same procedures apply to information provided by other interested groups (e.g., NIH, CDC, independent university investigators, consumer groups, and competitors) who may appear before the committees.

CBER also supplies the committee with information concerning the center's evaluation and analysis of the relevant IND or PLA. Data and information submitted by parties other than the product sponsor also may be provided.

While rigorous statistical criteria are applied in most cases, the committees may attempt to make benefit/risk assessments when the data are statistically inadequate or inconclusive. The committees are not asked to consider anecdotal statements or other unconfirmed or unconfirmable information, although such information is sometimes presented at the meetings.

According to CBER policy, the participants in the committee process—the committee members, CBER, and the product sponsor—should not be presented with, or asked to immediately comment on, new data or information. Such impromptu evaluations would be unscientific and not conducive to objective assessment.

There are, of course, instances in which newly developed information is critically important and becomes available at the last moment. Under these circumstances, the information will be made available, but only with the understanding that: (1) the participants will be alerted to the existence of the data prior to the meeting, even if it is the evening before; and (2) there is an increased probability that the committee will not be able to reach a conclusion on such information at that meeting.

Regardless of its source, information submitted for committee consideration must be channeled through the DFO. If relevant to the issues being considered by the committee, the information will be given to the members. If, in the center's judgment, the information is not relevant or is too voluminous, its appropriateness will be discussed with the submitter.

Committee Findings and Recommendations

In all cases, committee findings and recommendations to the FDA are advisory. The agency retains the authority and responsibility to act on such advice and make final regulatory decisions.

Committees are asked to make their recommendations as specific as possible, and to address fundamental issues such as product safety and effectiveness, benefit/risk assessments, the need for additional studies or information, the design and conduct of such studies, and the need for surveillance or other post-marketing actions. More detailed issues (e.g., product formulation, route of administration, potency, dosage schedule, stability, and labeling) are addressed as appropriate. Ideally, these issues should be considered within the context of the basic issue of product safety and effectiveness.

Specificity in committee recommendations is required by administrative law. Congress has empowered the Secretary of Health and Human Services and the Commissioner of Food and Drugs to make certain regulatory decisions on the basis of an administrative record consisting of an application's evidence of product safety, effectiveness and quality, the standards used in weighing the evidence, and the manner in which the standards were applied.

Therefore, it is essential that the FDA maintain accurate and complete records of advisory committee proceedings showing how and why the committees acted. Without this supporting documentation from the experts, the decision of the secretary or the commissioner can be challenged. In the final analysis, the agency must not only reach the right conclusion, but it must do so for the right reasons, all of which must be on record.

Accordingly, committees must base their advice on full knowledge of the topic under consideration and must provide this advice in as much detail as possible. If new and relevant information becomes available subsequent to a committee's recommendations, the committee can, and frequently does, reconvene to address the issue. It is not unusual for a CBER advisory committee to revisit a particular product or difficult issue at a number of meetings before final agency action.

There have been few instances in which CBER has clearly disagreed with committee recommendations. Disagreement can arise, however, when the agency receives additional or conflicting information subsequent to a committee's recommendation. CBER might also dissent if the committee deviated from agency regulations or policy in making its recommendations. For example, a committee might advise CBER that it believes a product to be effective on the basis of a single clinical trial lacking sufficient controls and statistical power, both of which are required for pivotal trials. Also, the agency might agree with a committee's scientific assessment of a product's safety and effectiveness as demonstrated in clinical trials but disagree with a committee's approval recommendation because of related issues that had not come before the committee (e.g., problems regarding establishment inspections, lot release, or labeling).

It should be noted that committee recommendations sometimes are conditional—that is, a committee's finding may be that the product appears to be safe and effective on the basis of available data but that the agency should request additional studies and information. Under such circumstances, it may take the product sponsor and/or CBER considerable time to implement a committee's tentative finding.

There is considerable speculation on how the FDA's use of advisory committees affects the approval time of new biological products. Data are sparse and inconclusive on this matter, however.

Dissemination of Committee Findings

Accounts of committee discussions and actions are accessible in several ways. Transcripts and summary minutes of open committee meetings are available from the FDA. Representatives of the national media and trade newsletters attend the meetings and report on committee findings. Also, professional journals frequently summarize committee deliberations.

If a committee's advice is favorable to a product sponsor or, conversely, to critics or opponents of a product, these groups frequently will disseminate their own press statements. If the issue is of wide interest, or is particularly sensitive or important, DHHS or the agency may also release a statement.

When committee deliberations are conducted in closed sessions to protect proprietary and other confidential information, the product's sponsor, investigators, and others with a vested interest generally will have participated in the closed deliberations. CBER informs these individuals—and others when appropriate—of the committee's deliberations.

Preparation By Participants at Advisory Committee Meetings

Much has been spoken and written on preparing for FDA advisory committee meetings. The essential advice can be condensed into seven specific recommendations:

1. Understand the process.
2. Be well prepared.
3. Understand and address the issues before the committee.
4. Provide concise but comprehensive pre-meeting briefing material with sufficient lead-time for study by members.
5. Develop presentations that are objective and consistent with good science.
6. Listen.
7. View the committee process as collegial and professional, not adversarial.

Finally, administrative remedies are available for individuals or companies that believe that the FDA or a committee is not complying with any portion of the regulations or the Federal Advisory Committee Act. Written petitions concerning past actions must be submitted within 30 days after the action to which an individual is objecting. Written petitions concerning future actions also are allowed and, if the objection pertains to an imminent or occurring action which could not have been anticipated, the matter may be handled by an oral petition. FDA regulations provide a detailed discussion of these remedies.

This chapter reflects the author's assessment of CBER's biologics advisory committees and is not intended to represent the official position of the FDA.

Chapter 12
Postlicensure Requirements

by Mark Mathieu
PAREXEL International Corporation

Similar to most other aspects of CBER's regulation of biologics, postlicensing requirements facing biological licensees were under revision during 1993. Adverse experience reporting, lot release, and advertising and promotional labeling preclearance policies were all being revised or clarified as of this writing. This chapter focuses on the revisions and clarifications as they will affect existing requirements.

Postlicensing requirements can be grouped into four general areas:

- adverse experience reporting requirements;

- lot release requirements;

- current Good Manufacturing Practice (cGMP) requirements; and

- general reporting requirements.

Adverse Experience Reporting for Licensed Biological Products

On March 29, 1990, the FDA proposed its first adverse experience reporting requirements for licensed manufacturers of biological products. Since the agency had never before proposed formal requirements for biologics, manufacturers reported on a voluntary basis under adverse experience reporting requirements for drugs. The 1990 proposal essentially reiterated the reporting requirements applicable to drugs under 21 CFR 314.80.

As of this writing, the final regulation (§ 600.80 Adverse Experience Reporting for Licensed Biological Products) was about to be published. A related guideline was to be made available with the final regulation.

With that final rule, the FDA was to publish a notice proposing revisions to adverse experience reporting for both human drug and licensed biological products. The purpose of the proposed revisions is to maintain consistency in drug and biologic postmarketing adverse experience reporting requirements and to incorporate several new initiatives into the regulations. On June 3, 1993, for example, the FDA unveiled the MedWatch Program, which is designed to encourage health-care professionals to report serious adverse events. In its upcoming regulatory proposal, the FDA will also introduce provisions designed to encourage international harmonization. This chapter emphasizes these initiatives as they will affect postlicensing adverse experience reporting for biological manufacturers.

Defining Adverse Experience   The adverse experience reporting area has very much its own language. In fact, reporting requirements are directly linked to criteria outlined in the definitions of three important terms: "adverse experience," "unexpected adverse experience," and "serious adverse experience."

An adverse experience is any adverse event coincident with a biologic's use without regard to the event's causality. Such experiences include:

- an adverse event occurring in the course of a product's use in professional practice;

- an adverse event occurring from overdose;

- an adverse event occurring from drug withdrawal; and

- any significant failure in the expected pharmacological action.

An unexpected event is an adverse experience that is not listed in the current labeling for the product. This includes events that may be symptomatically and pathophysiologically related to an event listed in the labeling or events that are of greater severity or specificity.

Under CBER's upcoming proposed regulation, the definition of "serious event" would be revised so that it is compatible with international definitions and with the definition outlined in the MedWatch Program. Therefore, a serious event whose outcome can be attributed to the adverse experience would be defined as follows:

> An adverse experience occurring at any dose that is fatal or life-threatening, results in persistent or significant disability/incapacity, requires or prolongs inpatient hospitalization, necessitates medical or surgical intervention to preclude impairment of a body function or permanent damage to a body structure, or is a congenital anomaly.

The phrase "occurring at any dose" would be added to provide for both labeled and unlabeled doses, including overdoses. "Cancer" and "overdose" would be dropped from the original definition.

The phrase "necessitates medical or surgical intervention to preclude impairment of a body function or permanent damage to a body structure" is not intended to be used to define an outcome as "serious" when a drug is used to treat the event. This phrase would be added for consistency with the definition of "serious" that was proposed by the International Conference on Harmonization of Technical Requirements for Registration of Pharmaceuticals for Human Use (ICH). The ICH promotes the international harmonization of technical requirements for pharmaceutical product registration among the United States, Japan, and the European Community.

Adverse Experience Reporting Requirements   Licensed manufacturers and any person whose name appears on the label of a licensed biological product as its manufacturer, packager, or distributor have reporting responsibilities. CBER's final regulations will specify requirements for four types of adverse experience reports:

- 15-day alert reports;

- periodic adverse experience reports;

- increased frequency reports; and

- distribution and disposition reports.

15-Day Alert Reports.   All adverse experiences that are both serious and unexpected must be reported by manufacturers within 15 days of their becoming aware of the event.

MED**W**ATCH

THE FDA MEDICAL PRODUCTS REPORTING PROGRAM

For use by user-facilities, distributors and manufacturers for MANDATORY reporting.

Page ____ of ____

Form Approved: OMB No. 0910-0291 Expires 12/31/94
See OMB statement on reverse

| Mfr. report # |
| U.F. Dist. report # |
| FDA Use Only |

A. Patient Information

1. Patient identifier

In confidence

2. Age at time of event:

or

Date of birth:

3. Sex
☐ female
☐ male

4. Weight
_____ lbs
or
_____ kgs

B. Adverse event or product problem

1. ☐ Adverse event and/or ☐ Product problem (e.g., defects/malfunctions)

2. Outcomes attributed to adverse event (check all that apply)
☐ death _____ (mo. day yr.)
☐ life-threatening
☐ hospitalization - initial or prolonged
☐ disability
☐ congenital anomaly
☐ required intervention to prevent permanent impairment/damage
☐ other _____

3. Date of event _____ (mo. day yr.)

4. Date of this report _____ (mo. day yr.)

5. Describe event or problem

6. Relevant tests/laboratory data, including dates

7. Other relevant history, including preexisting medical conditions (e.g., allergies race, pregnancy, smoking and alcohol use, hepatic/renal dysfunction, etc.)

C. Suspect medication(s)

1. Name (give labeled strength & mfr/labeler, if known)

#1.

#2

2. Dose, frequency & route used
#1.
#2

3. Therapy dates (if known, give duration)
#1.
#2

4. Diagnosis for use (indication)
#1.
#2

5. Event abated after use stopped or dose reduced
#1.☐ yes ☐ no☐ doesn't apply
#2 ☐ yes ☐ no☐ doesn't apply

6. Lot # (if known)
#1.
#2

7. Exp. date (if known)
#1.
#2

8. Event reappeared after reintroduction
#1.☐ yes ☐ no☐ doesn't apply
#2 ☐ yes ☐ no☐ doesn't apply

9. NDC # - for product problems only (if known)

10. Concomitant medical products and therapy dates (exclude treatment of event)

G. All manufacturers

1. Contact offices-names/address(& mfnng site for devices)

2. Phone number

3. Report source (check all that apply)
☐ foreign
☐ study
☐ literature
☐ consumer
☐ health professional
☐ user facility
☐ company representative
☐ distributor
☐ other

4. Date received by manufacturer

5.
(A) NDA #
IND #
PLA #
Pre-1938 ☐ yes
OTC ☐ yes
product

6. If IND, protocol #

7. Type of report (check all that apply)
☐ 5-day ☐ 15-day
☐ 10-day ☐ periodic
☐ initial ☐ follow-up #

8. Adverse event term(s)

9. Mfr.report number

E. Initial reporter

1. Name, address & phone #

2. Health professional?
☐ yes ☐ no

3. Occupation

4. Initial reporter also sent report to FDA
☐ yes ☐ no ☐ unk

FDA

FDA Form 3500A (6/93)

Submission of a report does not constitute an admission that medical personnel, user facility, distributor, manufacturer or product caused or contributed to the event.

On June 3, 1993, the MedWatch Program introduced a new reporting form for manufacturers, FDA Form 3500A, which incorporates the reporting of product problems and adverse experiences for biologics, devices, drugs and other products regulated by the agency (see sample form above). The new form replaces Form FDA 1639, which will no longer be accepted after December 3, 1993.

The new form is designed to enable the use of optical character recognition (OCR) and form removal technology to extract the information from the form and transfer the data into the FDA database without manual data entry. The new form is double-sided to enable the reporting of device adverse events and product problem reports from manufacturers, user facilities, and distributors. If a licensed manufacturer that uses computer-generated forms is reporting an event associated with a human drug or a biological product, FDA Form 3500A may be modified by replacing Section D (which contains device product data elements) with Section G (which includes manufacturer information). This modification results in a one-sided form that meets all of the requirements for reporting adverse events associated with a biological product or human drug.

To further promote international harmonization, adverse experience reports originating outside of the United States may be submitted on a form developed by the Council for International Organizations of Medical Sciences (CIOMS). An independent body located in the World Health Organization's headquarters in Geneva, Switzerland, CIOMS provides a forum for manufacturers and regulators to develop and test uniform approaches and formats for adverse experience reporting. The use of a CIOMS form enables a manufacturer to submit one adverse experience report form to all participating national regulatory agencies.

Periodic Adverse Experience Reports. Under CBER's upcoming proposed regulations, periodic adverse experience reporting requirements would be revised to conform to CIOMS II recommendations. The CIOMS II working party's goal was to develop a reporting format and frequency that would enable a manufacturer to prepare a single report that would satisfy requirements of national regulatory agencies throughout the world. The elements of the new periodic reporting requirements include:

• A six-month periodic reporting cycle.

- Submission of a periodic report within 45 days of the date on which the product was first licensed anywhere in the world. This is referred to as the international birth date.

- A "core data sheet" containing all relevant safety information. The core data sheet enables a manufacturer to produce a periodic report acceptable to all participating countries.

- A report format that is based on CIOMS II and that includes information on patient exposure and worldwide regulatory decisions concerning marketing, regulatory, or manufacturer actions taken for safety reasons.

Increased Frequency Reports. Licensed manufacturers must periodically review the frequency with which serious expected adverse experiences are reported. If a manufacturer determines that a particular event is being reported with substantially increased frequency compared to previous reporting intervals, the company must submit an increased frequency report within 15 days. The report must be submitted in narrative form, and must describe the method of the analysis, the time period upon which the analysis is based, and the interpretation of the results.

Licensed manufacturers should conduct periodic reviews at least as often as the periodic reporting cycle. To maintain consistency with the reporting cycle for periodic reports under the CIOMS II format, the FDA's proposed revisions establish a six-month periodic cycle.

Distribution and Disposition Reports. CBER's final regulations on adverse experience reporting establishes a new distribution and disposition report specific to licensed biological products. The report must include the product name, the bulk lot and fill lot numbers for the total number of dosage units of each strength or potency distributed, the quantities distributed for domestic and foreign use, and the amount of bulk lot remaining. The submission of distribution and disposition reports is based on the CIOMS II periodic reporting cycle. The disclosure of financial or pricing data is not required.

Adverse Experience Reporting for Vaccine Products The National Childhood Vaccine Injury Act of 1986 requires physicians and other health-care professionals who administer vaccines to maintain permanent records. These records must include the date of vaccine administration, the vaccine manufacturer, the product's lot number, and the name, address, and title of the person administering the vaccine. The statute also established a vaccine injury table (VIT) that specifies occurrence intervals for reportable events following vaccination with specific vaccines.

The Vaccine Adverse Event Reporting System (VAERS) was launched in November 1990 to provide a central database for the collection of vaccine-related adverse experience reports. Reports submitted directly to VAERS by health-care professionals represent a significant portion of the reports in the database. The provisions of CBER's final regulation are not expected to significantly increase the percentage of reports submitted by manufacturers.

The final rule requires licensed manufacturers to report adverse events involving vaccines. Aside from the fact that vaccine reports must be made on a VAERS-1 form (see sample form below), requirements for vaccine and non-vaccine adverse experience reporting are identical. It is also important to note that confidential information concerning the report and the patient are subject to the Centers for Disease Control and Prevention Privacy Act.

Lot Release

CBER's focus on product manufacturing is apparent in several postlicensure requirements. After licensure, a biological establishment must supervise and control the manufacturing process to ensure, among other things, that contaminants are not introduced during production and that there is lot-to-lot consistency in the quality of the licensed product. CBER's lot release program is one mechanism through which CBER ensures that licensees fulfill these duties.

Administered by CBER's Division of Product Quality Control, the lot release program is designed to ensure the consistency of a licensee's manufacturing processes, and to ensure that licensed products continue to meet regulatory standards specified in the PLA and federal regulations. To obtain lot release authorization, licensees must submit representative lot samples and a lot release protocol—a summary of the manufacturer's test results on the lot. Laboratories within the Division of Product Quality Control and other CBER divisions review the results and may conduct their own testing to confirm the manufacturer's results. Manufacturers may not distribute production lots until they receive CBER's authorization, formally called an "official release."

VACCINE ADVERSE EVENT REPORTING SYSTEM		*For CDC/FDA Use Only*
VAERS 24 Hour Toll-free information line 1-800-822-7967 P.O. Box 1100, Rockville, MD 20849-1100 PATIENT IDENTITY KEPT CONFIDENTIAL		VAERS Number _____ Date Received _____

Patient Name:	Vaccine administered by (Name):	Form completed by (Name):
Last First M.I. Address	Responsible Physician _____ Facility Name/Address	Relation ☐ Vaccine Provider ☐ Patient/Parent to Patient ☐ Manufacturer ☐ Other Address (*if different from patient or provider*)
City State Zip Telephone no. (____) _____	City State Zip Telephone no. (____) _____	City State Zip Telephone no. (____) _____

1. State	2. County where administered	3. Date of Birth mm dd yy	4. Patient age	5. Sex ☐ M ☐ F	6. Date form completed mm dd yy

7. Describe adverse event(s) (symptoms, signs, time course) and treatment, if any	8. Check all appropriate: ☐ Patient died (date ___/___/___ mm dd yy) ☐ Life threatening illness ☐ Required emergency room/doctor visit ☐ Required hospitalization (_____ days) ☐ Resulted in prolongation of hospitalization ☐ Resulted in permanent disability ☐ None of the above

9. Patient recovered ☐ YES ☐ NO ☐ UNKNOWN	10. Date of Vaccination ___/___/___ mm dd yy Time _____ AM PM	11. Adverse event onset ___/___/___ mm dd yy Time _____ AM PM
12. Relevant diagnostic tests/laboratory data		

13. Enter all vaccines given on date listed in no. 10				
Vaccine (type)	Manufacturer	Lot number	Route/Site	No.Previous doses
a.				
b.				
c.				
d.				

14. Any other vaccines within 4 weeks of date listed in no. 10				
Vaccine (type)	Manufacturer	Lot number	Route/Site	No. Previous doses Date given
a.				
b.				

15. Vaccinated at: ☐ Private doctor's office/hospital ☐ Military clinic/hospital ☐ Public health clinic/hospital ☐ Other/unknown	16. Vaccine purchased with: ☐ Private funds ☐ Military funds ☐ Public funds ☐ Other/unknown	17. Other medications
18. Illness at time of vaccination (specify)	19. Pre-existing physician-diagnosed allergies, birth defects, medical conditions (specify)	

20. Have you reported this adverse event previously? ☐ No ☐ To doctor ☐ To health department ☐ To manufacturer	*Only for children 5 and under*	
	22. Birth weight _____ lb. _____ oz.	23. No. of brothers and sisters
21. Adverse event following prior vaccination (check all applicable, specify)	*Only for reports submitted by manufacturer/immunization project*	
Adverse Event Onset Age Type Vaccine Dose no. in series ☐ In patient ☐ In brother ☐ or sister	24. Mfr./imm. proj. report no.	25. Date received by mfr./imm. proj.
	26. 15 day report? ☐ Yes ☐ No	27. Report type ☐ Initial ☐ Follow-Up

Health care providers and manufacturers are required by law (42 USC 300aa-25) to report reactions to vaccines listed in the Vaccine Injury Table. Reports for reactions to other vaccines are voluntary except when required as a condition of immunization grant awards

Form VAERS-1

The types and quantity of testing that CBER conducts on product samples depends on several factors, including the nature of the product and the perceived stability of the licensee's manufacturing operation. Manufacturers of biotechnology-derived injectable biologics, for example, must submit samples of both the purified bulk product and the final container product. On the purified bulk product sample, CBER laboratories might conduct testing regarding protein characterization, identity, impurities, and viral, nucleic acid, antigen, microbial, and endotoxin contamination. The final product might be tested for purity, potency, sterility, general safety, pyrogenicity, identity, moisture, and preservatives.

Although the agency requires it for most biological products, the submission of samples for lot release is not an automatic postlicensure requirement. CBER must notify a licensee that its product is subject to lot release, something the center typically does at product licensure. According to federal regulations, "Upon notification by the Director, Center for Biologics Evaluation and Research, a manufacturer shall not distribute a lot of a product until the lot is released by the Director, Center for Biologics Evaluation and Research: *Provided*, That the Director, Center for Biologics Evaluation and Research, shall not issue such notification except when deemed necessary for the safety, purity, or potency of the product."

Alternatives To Lot Release In a July 20, 1993 guidance document, CBER outlined how biologics manufacturers may apply for exemptions from lot release requirements. CBER officials point out that manufacturers of a few products have been exempted from these requirements and that the document simply outlines what has been a long-standing center policy.

According to the July 1993 statement, "Current technology combined with the experience derived from years of product-specific inspections and testing in CBER laboratories has demonstrated that, for some biological products, alternatives to requiring a CBER release action for every lot provide adequate control to ensure continued safety, purity, and potency (including effectiveness)."

Under the provisions of the guidance document, manufacturers that have "documented an acceptable history of lot release and control of the manufacturing facility" may submit a PLA amendment requesting the approval of the lot release alternatives. Standards regarding what comprises an "acceptable lot release history will vary according to the product and the complexities of the manufacturing process," says the agency.

In the notice, CBER does not suggest what these "alternatives" might be, offering only that the licensee must demonstrate that "the alternative approach does not compromise the safety, purity, and potency of the biological product." Assuming this requirement is met, "CBER may consider whether there is a need for manufacturers to submit samples and protocols at specific intervals (e.g., quarterly) for surveillance purposes."

CBER identifies five types of data that should be submitted in PLA amendments proposing an alternative to lot release. These data, which must cover "an adequate period of time and a sufficient number of production lots," should include:

"(1) A well-organized table containing a testing summary of all lots manufactured, including lots manufactured in support of licensure. This testing history should include both lots submitted to CBER for release action and lots or batches rejected during in-process, bulk, or final testing at the manufacturing establishment.

(2) A summary of the disposition of the above lots, including the reason a final lot was not submitted to CBER for release or an in-process, or bulk lot or batch was rejected.

(3) A summary listing all product complaints which include, but are not limited to, presence of labeling errors, decreased potency, contamination, particulate matter, adverse reactions, and defect reports. The action taken by the manufacturer for each identified production lot or batch should be described.

(4) A listing of any lot(s) which was subject to recall or market corrective action following distribution.

(5) A description of any major process change, including when the process change was implemented and a list of lots manufactured using the new procedure."

In reviewing these data, CBER will give primary consideration to the licensee's manufacturing history and the likelihood that the lot release alternative will compromise the product's safety, purity, potency, and stability. "Among the factors that CBER assesses in determining whether to approve such amendment requests are conformance to licensed manufacturing procedures and the ability of the manufacturer to consistently demonstrate product safety, purity, potency, and stability," the center states. "In addition, there should be a history of FDA establishment inspections that have shown compliance with applicable regulations during the period covered. The period considered may vary by product, because the number of lots produced in a given time may vary, as may the extent to which lot release procedures are viewed as important for ongoing assurance of safety and efficacy. CBER recognizes that the need for submission of lot release protocols and/or samples may be greater for some products than others, e.g., products where maintenance of consistent specification from lot-to-lot is difficult and/or where insufficient correlation is available between measurement of potency and biological activity. The experience reflected in both the number of lots produced and the period of production is important to assess the potential value of the lot release procedures for a particular product or product class."

CBER is clear regarding its authority to reimpose lot release requirements once alternatives have been approved. In response to a major change in a manufacturing process or establishment or the failure of a product surveillance sample to meet an established specification, the center may reimpose lot release requirements.

cGMP for Biological Products

Since the early 1960s, federal regulations have required firms producing human drugs and biologics to operate their facilities under current Good Manufacturing Practice (cGMP) standards. cGMP regulations specify general standards for manufacturing facilities and their production controls and operations to ensure that products made in these facilities meet relevant standards of identity, strength, quality, and purity.

When cGMP first became a requirement, the FDA had envisioned developing and publishing several different cGMPs, with separate standards for finished dosage forms, bulk ingredients, and other classes of products. To date, however, the FDA has published only one set of general cGMP standards—current Good

Manufacturing Practice for Finished Pharmaceuticals. There are, however, several cGMP regulations that establish additional standards for biological and biotechnology products (see discussion below).

Given that they accommodate the diverse manufacturing and control processes used in the manufacture of drug and biological products, the "core" cGMP regulations feature broad provisions that define general standards without offering specifics on how these may be fulfilled. The regulations address ten areas:

- organization and personnel;

- buildings and facilities;

- equipment;

- control of components and biological product containers and closures;

- production and process controls;

- packaging and labeling controls;

- holding and distribution;

- laboratory controls;

- records and reports; and

- returned and salvaged biological products.

Organization and Personnel   Several principal cGMP requirements focus on the quality control unit. Each manufacturing facility must have a quality control unit to ensure compliance with cGMP. According to federal regulations, the quality control unit assumes "the responsibility and authority to approve or reject all components, [biological] product containers, closures, in-process materials, packaging materials, labeling, and [biological] products and the authority to review production records to assure that no errors have occurred or, if errors have occurred, that they have been fully investigated." In addition, the one-or-more-person quality control unit is responsible for approving or rejecting all procedures or specifications affecting the identity, strength, quality, and purity of the biological product.

Obviously, the professionals responsible for performing, supervising, or consulting on the manufacture, processing, packing, or holding of a biologic must be adequate in number, and must be sufficiently qualified by education, training, and experience to carry out their respective tasks. Facility staffers must be trained not only in their specific responsibilities, but also in cGMP.

Buildings and Facilities   cGMP building and facility requirements are designed to ensure that any structures used to manufacture, process, pack, or hold a product are suitable in size, construction, and location to allow proper cleaning, maintenance, and operation. These requirements call for the separation of several plant operations—in-process materials storage, packaging and labeling operations, control and laboratory operations, and seven other operations must have distinct work areas to reduce the possibility of cross-contamination and other mishaps. Specific requirements for lighting, ventilation, heating and cooling systems, plumbing, sanitation, and maintenance are also provided.

Equipment   cGMP equipment requirements focus largely on equipment design, size, location, and maintenance. To ensure that product attributes are not adversely affected, surfaces that contact components, in-process materials, or biological products must not be reactive, additive, or absorptive. Lubricants and other substances required for equipment operations must not cause product contamination.

At predetermined intervals, all utensils and equipment must be cleaned, maintained, and sanitized according to specific written procedures. Filters and automatic, mechanical, and electronic equipment (including computers) must meet special requirements.

Components and Biological Product Containers and Closures Controls   A facility must maintain detailed written procedures for the receipt, identification, storage, handling, sampling, testing, and approval or rejection of components and product containers and closures. Upon receipt, these materials must first be inspected visually for appropriate labeling and contents, container damage, broken seals, and contamination. Before using the components, a facility must draw a sample from each lot, and test the samples for identity and conformity to purity, strength, and quality specifications.

Product containers and closures must be tested for conformance with applicable written requirements.

Production and Process Controls   Manufacturing plants must maintain written procedures for production and process controls designed to assure that the biological products have the identity, strength, quality, and purity they are represented to possess. Special cGMP requirements exist for the charge-in of components, yield calculations, the identification of compounding and storage containers, processing lines, and major equipment used in the production of product batches, the sampling and testing of in-process materials and biological products, and the limiting of production times.

Packaging and Labeling Controls   All packaging and labeling materials must be sampled, examined, or tested before their use. Documented procedures must be established for the receipt, identification, storage, handling, sampling, examination, and testing of labeling and packaging materials. The cGMP regulations also specify requirements for labeling issuances, packaging, and labeling operations control and inspection, product inspection, and expiration dating.

Holding and Distribution Requirements   Facilities must maintain detailed written procedures describing the warehousing operations (including quarantine and special storage procedures) and distribution methods used for biological products.

Laboratory Controls   Each organizational unit within a firm must draft procedures for control mechanisms such as specifications, standards, sampling plans, and testing procedures designed to assure that components, product containers, closures, in-process materials, labeling, and biological products conform to appropriate standards of identity, strength, quality, and purity. Laboratory controls, which must be reviewed and approved by the quality control unit, include:

- confirmation, through documented sampling and testing procedures, that each shipment lot of components, product containers, closures, and labeling conforms with relevant specifications;

- confirmation, through sampling and testing procedures, that in-process materials conform to written specifications;

- confirmation that the laboratory is complying with written descriptions of product sampling procedures and specifications; and

- confirmation that instruments, apparatus, gauges, and recording devices have been calibrated at suitable intervals according to written procedures that provide specific directions, schedules, limits for accuracy and precision, and provisions for remedial action in the event accuracy and/or precision limits are not met.

The facility must develop a written program for stability testing procedures. The results of this testing are used to determine appropriate storage conditions and expiration dates for each product.

Also included in the laboratory controls subpart of the cGMP regulations are requirements for: (1) the sampling, testing, and release of product batches; (2) reserve sample retention, laboratory test animals, and special testing for sterile, pyrogen-free, ophthalmic, and controlled-release drugs; and (3) testing for penicillin contamination.

Records and Reports   Facilities must retain records for all product components, product containers, closures, and labeling for at least one year after the expiration date. cGMP regulations also specify requirements for master and batch production records, laboratory records, distribution records, and complaint files.

Returned and Salvaged Biological Products   Any returned biological products must be identified and held. Such products must be destroyed if there is any doubt about their safety, identity, strength, quality, or purity. Although returned drugs are sometimes reprocessed, the reprocessing of returned biological products is rare.

Additional Standards for Biological Products

Given the broad nature of the core cGMP regulations and the many special concerns regarding biologic manufacturing, CBER has developed a series of regulations that establish additional standards for biological and biotechnology product testing and manufacturing. In general terms, these regulations reflect CBER's desire for biologics manufacturers to exert greater control over their manufacturing processes and the physical environment to which biological products are exposed:

- 21 CFR Part 600 - Biological Products: General;

- 21 CFR Part 601 - Licensing;

- 21 CFR Part 606 - Current Good Manufacturing Practice for Blood and Blood Components;

- 21 CFR Part 607 - Establishment Registration and Product Listing for Manufacturers of Human Blood and Blood Products;

- 21 CFR Part 610 - General Biological Products Standards;

- 21 CFR Part 620 - Additional Standards for Bacterial Products;

- 21 CFR Part 630 - Additional Standards for Viral Products;

- 21 CFR Part 640 - Additional Standards for Human Blood and Blood Products;

- 21 CFR Part 650 - Additional Standards for Diagnostic Substances for Dermal Tests;

- 21 CFR Part 660 - Additional Standards for Diagnostic Substances for Laboratory Tests; and

- 21 CFR Part 680 - Additional Standards for Miscellaneous Products (allergenic products, trivalent organic arsenicals, and blood group substances).

To supplement these regulations, the FDA has also published guidelines applicable to drug and biologic cGMP, including the *Guideline on Sterile Drug Products Produced by Aseptic Processing* and the *Guideline on General Principles of Process Validation*. In addition, CBER has published several points-to-consider documents that provide insights into generally accepted FDA standards for the testing and production of certain biotechnology products, including monoclonal antibodies, products made through recombinant DNA technology, and interferon (see Appendix 1).

Lastly, several FDA inspectional guides developed for agency inspectors can provide manufacturers with valuable insights into cGMP requirements, inspections, and problem areas. Among these are the FDA's *Biotechnology*

Inspections Guide (November 1991), the *Inspection Operations Manual,* and the *Compliance Program Guidance Manual.*

(Editor's note: For information on CBER's prelicensing and periodic inspections of manufacturing facilities, see Chapter 9.)

General Reporting Requirements

Biologics licensees face several additional postlicensure requirements in addition to the lot release, cGMP, and adverse experience reporting requirements described above. The FDA has three additional postmarketing requirements:

- reporting changes in manufacturing;

- reporting errors; and

- reporting advertising and promotional labeling changes.

Reporting Manufacturing Changes   Given the orientation of biologics regulation to manufacturing issues, it is not surprising that licensees face relatively stringent postmarketing reporting requirements for changes regarding product manufacturing. Failure to meet these requirements can also carry a steep price: license revocation.

Sponsors must report to CBER important proposed changes in facility location, equipment, management and responsible personnel, and manufacturing methods at least 30 days prior to implementation. Since even minor changes in the manufacturing process may have significant effects on the product, all changes must be reported to CBER prior to implementation. Some changes must be approved by CBER before implementation, including changes to manufacturing methods, labeling, and equivalent methods and processes. Licensees must submit proposed changes as amendments to the relevant PLA or ELA.

CBER's policy on manufacturing changes has been a point of contention for some time, with industry arguing that the policy is overly stringent. Since the policy requires that companies report such minor changes as the replacement of control or measuring devices and standard biochemical processing equipment, industry representatives claim that CBER's policy creates unnecessary work for CBER and industry, and delays beneficial manufacturing process changes.

259

In mid-1993, CBER was developing a new policy that is expected to clarify the center's postmarketing reporting requirements. The policy is also expected to ease reporting requirements regarding manufacturing changes.

Reporting of Errors According to federal regulations, licensees must promptly notify CBER's Office of Compliance "... of errors or accidents in the manufacture of products that may affect the safety, purity, or potency of any product." Such reporting can alert CBER to manufacturing- or facility-related problems, and information that will allow the center to consider whether a product recall or other regulatory action is necessary.

There is no standard form for sponsor reporting of errors. Information describing the error should be detailed in a letter accompanied by any necessary attachments.

Reporting Advertising and Promotional Labeling Changes As part of its reorganization, CBER created a unit dedicated to reviewing advertising and promotional labeling for biologics—the Advertising, Promotion, and Labeling Staff (APLS)—within the Office of Establishment Licensing and Product Surveillance. The group was formed in an effort to attain consistency in the review of promotional materials and to enable product reviewers to concentrate on PLAs and amendments. Reviewers within the product offices will continue to be responsible for evaluating and preclearing all changes to the package insert, container label(s), and carton label(s).

Among the unit's first major accomplishments was the release of *The Advertising and Promotional Labeling Staff Procedural Guide* in August 1993. According to the procedural guide, "In an effort to reduce the burden of reporting, [the] Center for Biologics Evaluation and Research (CBER) is providing guidance on its current interpretation of the regulations in effect for advertising and promotional labeling. CBER has determined that preapproval of promotional labeling is required only for any biological product for which a license is pending, any licensed biological products for which there is a significant amendment, and for 'newly approved' products (the first 120 days following approval). It is requested that manufacturers continue to submit for review only the introductory advertising materials to be used in the first 120 days following approval."

Form Approved; OMB No. 0910-0039.
Expiration Date: January 31, 1992
See OMB statement on reverse.

DEPARTMENT OF HEALTH AND HUMAN SERVICES **PUBLIC HEALTH SERVICE** **FOOD AND DRUG ADMINISTRATION** CENTER FOR BIOLOGICS EVALUATION AND RESEARCH **TRANSMITTAL OF LABELS AND CIRCULARS** (Part I - Submission Of Draft And Preliminary Proof Labeling)	**LABEL REVIEW NO.**

NOTE: *No license may be granted unless this completed submittal form has been received (U.S. Public Health Service Act, Section 351; the Federal Food, Drug, and Cosmetic Act, Section 502; and Title 21 U.S. Code of Federal Regulations, Part 600).*

GENERAL INSTRUCTIONS

Type or print legibly in ink. Submit three copies of preliminary proofs and drafts. For revised labeling, indicate where changes have been made on the labeling copy. Assemble and staple each set, including attachments. The transmittal form must be dated and signed by the responsible head. Return both parts (1 and 2) of this form to the Food and Drug Administration, Center for Biologics Evaluation and Research (HFB-240), 8800 Rockville Pike, Bethesda, Maryland 20892.

MANU-FACTURER'S NAME	
NAME OF PRODUCT	

LABELING	**CHECK BELOW**	**TYPE SUBMITTED**	**REPLACE LABELING PREVIOUSLY REVIEWED** REVIEW NO. / DATED	**MANUFACTURER'S IDENTIFICATION NO.**	**LABELING REPRESENTS CHANGE IN:**
		A Container Label			☐ Dosage ☐ Manufacturing Method ☐ Contraindications, side effects, Precautions ☐ Arrangement ☐ Wording ☐ Other *(Specify)*
		B Package Label			
		C Circular			
		D Diluent			
		E Other *(Specify)*			

CHECK THE BOX PROVIDED IF THIS LABELING IS IN SUPPORT OF LICENSE APPLICATION OR AMENDMENT	CHECK HERE ➡	REFERENCE NUMBER

COMMENTS		

RESPONSIBLE HEAD	SIGNATURE	DATE

THE SPACE BELOW IS FOR REVIEW BY CENTER FOR BIOLOGICS EVALUATION AND RESEARCH

COMMENTS		

REVIEWED BY	SIGNATURE	DATE
RETURNED BY	SIGNATURE	DATE

FORM FDA 2567 (7/90)   PREVIOUS EDITION MAY BE USED

DEPARTMENT OF HEALTH AND HUMAN SERVICES PUBLIC HEALTH SERVICE FOOD AND DRUG ADMINISTRATION CENTER FOR BIOLOGICS EVALUATION AND RESEARCH **TRANSMITTAL OF LABELS AND CIRCULARS** *(Part II - Implementation Of Final Printed Labeling)*	LABEL REVIEW NO.

NOTE: *No license may be granted unless this completed submittal form has been received (U.S. Public Health Service Act, Section 351; the Federal Food, Drug, and Cosmetic Act, Section 502; and Title 21 U.S Code of Federal Regulations, Part 600).*

GENERAL INSTRUCTIONS

Draft or proof of labeling submitted under the above review number has been returned to you with this form. At the time final printed labeling is put into use, attach three complete sets of labeling to this form and complete the implementation information below. Return this form with final printed labeling to the Food and Drug Administration, Center for Biologics Evaluation and Research (HFB-240), 8800 Rockville Pike, Bethesda, Maryland 20892.

MANU- FACTURER'S NAME						
NAME OF PRODUCT						
LABELING	CHECK BELOW	TYPE SUBMITTED	REPLACE LABELING PREVIOUSLY REVIEWED REVIEW NO / DATED	MANUFACTURER'S IDENTIFICATION NO.	LABELING REPRESENTS CHANGE IN: ☐ Dosage ☐ Manufacturing Method ☐ Contraindications, side effects, Precautions ☐ Arrangement ☐ Wording ☐ Other *(Specify)*	
		A Container Label				
		B Package Label				
		C Circular				
		D Diluent				
		E Other *(Specify)*				

CHECK THE BOX PROVIDED IF THIS LABELING IS IN SUPPORT OF LICENSE APPLICATION OR AMENDMENT	CHECK ➡ HERE		REFERENCE NUMBER

IMPLEMEN- TATION	DATE ATTACHED LABELING WAS INITIALLY DISTRIBUTED WITH PRODUCT LOT NUMBER OF THE INITIAL LOT UTILIZING ATTACHED FINAL PRINTED LABELING

RESPONSIBLE HEAD	SIGNATURE	DATE

THE SPACE BELOW IS FOR REVIEW BY CENTER FOR BIOLOGICS EVALUATION AND RESEARCH

COMMENTS		
REVIEWED BY	SIGNATURE	DATE
RETURNED BY	SIGNATURE	DATE

FORM FDA 2567 (7/90)   PREVIOUS EDITION IS OBSOLETE

Previous CBER policy required that all promotional labeling (i.e., brochures, file cards, "Dear Doctor" letters, etc.) be precleared through the center before dissemination. This preclearance procedure applied to changes to previously reviewed and cleared materials as well. In the past, CBER also requested that manufacturers submit all advertising for review before publication or dissemination.

"Applicants should submit introductory advertising and promotional labeling ('launch' materials) in duplicate to [APLS] prior to the issuance of a biological product license and/or receipt of an approval letter," says the CBER procedural guide. "CBER will review all advertising and promotional labeling for products that have been licensed for more than 120 days on a selected surveillance basis. Applicants should continue to submit to APLS final copies of all advertising and promotional materials at the time of initial publication or distribution for inclusion in their product files."

CBER will continue to require that a completed Form FDA 2567, *Transmittal of Labels and Circulars* (see sample form above) accompany "all final printed, published or produced material and all media." The form should be used for all introductory advertising and promotional labeling, all materials following approval, and all sponsor requests for CBER review and comment.

Chapter 13
The FDA's Bioresearch Monitoring Program

by Steven Falter
Director, Division of Bioresearch Monitoring and Regulation
Center for Biologics Evaluation and Research
Food and Drug Administration

Because the FDA's regulatory decisions are based directly on research data, the agency has a vested interest in ensuring the accuracy and validity of non-clinical and clinical study results. Under CBER's Bioresearch Monitoring Program, agency inspectors visit laboratories and offices where these scientific data are developed and stored.

Specifically, the FDA conducts on-site inspections at clinical investigation sites, biologics sponsors, institutional review boards (IRB), and nonclinical laboratories to ensure: (1) that data submitted in product applications are reliable; and (2) that the rights and welfare of human subjects in clinical trials are protected. During these inspections, FDA investigators evaluate how well sponsors, monitors, clinical and nonclinical investigation site staff, and IRBs have fulfilled their respective responsibilities and commitments under GCPs, research protocols, and related standards and regulatory requirements.

CBER's Bioresearch Monitoring Program operates under similar—in some cases identical—regulations and policies to those used by CDER. Any differences between the two programs relate to the innate differences between the molecular entities under study. Biological products often are large molecules with complex structures and often are difficult to evaluate in the laboratory. Also, because there are no generic biologics for which bioequivalence/bioavailability studies are needed, CBER has no counterpart to CDER's program to monitor these studies.

A Brief History of the Bioresearch Monitoring Program

The origin of the FDA's authority to review clinical data and related records is the Food, Drug and Cosmetic (FD&C) Act. The law states that every person required to maintain records must, upon the FDA's request, allow access to clinical data for review and copying. For example, FDA regulations—specifically, Form FDA 1572, which clinical investigators sign before undertaking the study of an investigational drug—state "...the investigator will make such records available for inspection and copying."

Because of physicians' importance in the development and collection of clinical safety and effectiveness data, inspections of investigators are the core of the FDA's Bioresearch Monitoring Program. FDA inspections of clinical investigators began in 1962, although only three inspections were done by 1965. The agency expanded its efforts in the years following and established a four-person office to organize and conduct inspections. But because government authorities outside the FDA believed that only physicians should evaluate the practices of other physicians, inspectional activities were limited (i.e., only seven or eight inspections were conducted annually).

By 1972, however, the U.S. government had gained a new respect for the importance and abilities of FDA inspectors. As a result, the FDA initiated a survey of 162 commercially sponsored clinical investigators, 70 noncommercial clinical investigators, and 15 manufacturers. The results of this multi-year study, and increased FDA staff and budget, led to the founding of the agency's Bioresearch Monitoring Program in June 1977.

Through the late 1970s and early 1980s, the FDA's Bureau of Biologics maintained a small Bioresearch Monitoring Program. In 1982, the FDA reorganized and combined drug and biologics regulation, including the Bioresearch Monitoring Program, under a single unit called the Center for Drugs and Biologics.

A 1987 reorganization once again separated drug and biologics regulation. CBER, which was created to regulate biological products, established a Bioresearch Monitoring Branch to inspect and evaluate biologics research studies.

More recent events also stand to have fundamental effects on CBER's bioresearch monitoring initiatives. With the passage of the Prescription Drug User Fee Act of 1992, for example, CBER's bioresearch monitoring staff is likely to grow. Because CBER now faces PLA review deadlines, the center will use a portion of the fees collected to streamline bioresearch monitoring inspections to ensure that they do not present unnecessary delays in product approvals.

CBER's Bioresearch Monitoring Program consists of four separate compliance inspection programs, each designed to evaluate the activities of a key figure in the conduct of a scientific study. The programs are:

• Clinical Investigator Compliance Program;
• Sponsor/Monitor Compliance Program;
• Institutional Review Board (IRB) Compliance Program; and
• Nonclinical Laboratory Compliance Program.

Actually, CBER's Bioresearch Monitoring Branch manages only three of these four inspectional areas. CDER's Bioresearch Monitoring Branch manages the Nonclinical Laboratory Compliance Program for both drugs and biologics (see Chapter 3). This chapter profiles CBER's clinical investigator and sponsor/monitor compliance programs.

The Clinical Investigator Compliance Program

As stated above, inspections that target clinical investigators are the core of the FDA's bioresearch monitoring initiatives. CBER manages two clinical investigator inspection programs: the Data Audit Program and the "For Cause" Inspection Program.

Data Audit Program The Data Audit Program is a routine surveillance program that involves the inspection of clinical research sites. During these inspections, an FDA field inspector reviews the conduct of the study and performs a data audit. In evaluating the investigator's conduct of a clinical study, the inspector reviews several factors:

• what the investigator, each of his/her staffers, and others did during the study;

• the degree of delegation of authority;

267

- when specific aspects of the study were performed;

- how and where data were recorded;

- how the biological substance was stored and accounted for;

- the monitor's interaction with the physician; and

- evidence of IRB approval for studies performed.

CBER generally will initiate a data audit inspection of a clinical study if:

- the product under study is a significant new biologic or is being studied for a significant new indication;

- the clinical studies for a product present significant safety issues for study subjects, such as in studies of pediatric vaccines;

- a clinical study is viewed as key in determining the product's effectiveness; or

- CBER has accepted for review, or expects to receive, a PLA for a breakthrough product, such as an AIDS vaccine, that may be subject to accelerated approval.

In the typical routine audit, data submitted in an application are compared with all on-site records that should support their validity. On-site records to which an FDA inspector must be given access include a physician's office records, hospital records, and various laboratory reports. Records obtained prior to the initiation and following the completion of the study may be reviewed and copied.

When CBER has targeted key clinical studies for inspection, the center will thoroughly review the PLA and other information to determine whether there are aspects of the study that deserve close investigation. For example, the sponsor may report in the PLA fewer adverse experiences than CBER scientists might typically expect for the type of product under study. Although the difference may be related to a better safety profile for the product, CBER's investigation will ensure that the disparity in adverse experiences is not due to underreporting.

After completing this review, CBER staffers send what is called an "assignment package"—copies of a representative number of case reports, the protocol, and other pertinent information—to a field inspector, who then contacts the investigational site. Due to the nature of the physician's work and schedule, inspections are made by appointment. Therefore, investigators are alerted as to which study will be audited and are given the opportunity to locate and prepare all relevant records.

CBER's "For Cause" Inspection Program   Under some circumstances, CBER conducts inspections of clinical investigations for reasons other than assuring the accuracy and completeness of data submitted in PLAs. CBER initiates "for cause" inspections when it determines there is a need to investigate a particular issue, but that an overall audit of the records and data related to a clinical study is unnecessary. CBER would initiate a "for cause" inspection in response to:

- complaints from study subjects or other sources, such as the sponsor's competitors;

- evidence that an investigational product is being promoted or sold as a safe and effective product;

- evidence that the health and safety of study subjects are being jeopardized unduly, or the subjects are not being informed of their rights; or

- evidence that false statements may have been made or submitted to the FDA.

In most cases, CBER addresses issues related to the adequacy of data in routine data audits. Conversely, the center addresses questions and issues related to the safety and rights of study subjects, significant violations of the law, and complaint resolution in "for cause" inspections.

During "for cause" inspections, CBER may choose to bypass many of the records relevant to the study. However, the center will examine, in detail, the records and other evidence necessary to resolve the issues that triggered the inspection. At a later date, the same clinical investigation may become the subject of a routine data audit, at which time CBER will review the remaining records and data.

The Clinical Sponsor/Monitor Compliance Program

The Clinical Sponsor/Monitor Compliance Programs within both CBER and CDER are receiving increasing attention. In several clinical site inspections, agency investigators discovered significant problems, including chronic deviations from study protocols and unexplained differences between raw data and data submitted in premarketing applications.

Because of staffing limitations, CBER generally does not initiate inspections of sponsor/monitors without evidence—resulting from a clinical site inspection for example—that a sponsor or monitor has not fulfilled applicable regulatory responsibilities. Therefore, most CBER investigations of sponsors/monitors are "for cause" inspections.

The following profile outlines the full range of issues CBER might consider during sponsor/monitor inspections. In doing so, the profile identifies and describes the set of activities that might be undertaken during routine sponsor/monitor inspections. Depending on the nature of a suspected problem, CBER might focus on any one or more of these areas during the more frequent "for cause" inspections. If the problem appears to be that the clinical studies were, in general, poorly managed, most or all aspects of sponsor/monitor responsibilities may be addressed.

The goals of sponsor/monitor inspections are to determine: (1) if the sponsor and/or any employee or firm contracted to do so is monitoring a clinical investigation adequately; and (2) if the sponsor is fulfilling each of its requirements as outlined in federal regulations and guidelines.

Although the clinical sponsor and its activities are the primary focus of inspections conducted under this program, it is worth mentioning that two other entities, because they may assume some of the sponsor's responsibilities during a clinical study, may also be investigated by the FDA:

- *Clinical Monitors.* Monitors are individuals selected by either a sponsor or contract research organization to oversee the progress of the clinical investigation. The monitor may be an employee of the sponsor, a contract research organization, or a consultant.

• *Contract Research Organization (CRO).* CROs are organizations or corporations that enter into contractual agreements with sponsors to perform one or more of the obligations of sponsors during clinical trials. A commercial drug sponsor may delegate several of its responsibilities to a CRO, including the design of a protocol, the monitoring of clinical studies, the selection of investigators and study monitors, the evaluation of reports, and the preparation of materials to be submitted to the FDA.

After identifying a firm for inspection, CBER's Bioresearch Monitoring Branch forwards an inspection assignment to the relevant FDA regional office—the office that oversees the district in which the sponsor/monitor's headquarters is located. The assignment package provides the FDA field inspector with the name and address of the sponsor/monitor, instructions on which studies are to be investigated, and information on what aspects of the studies should be emphasized during the inspection.

Sponsor inspections are conducted at the company's headquarters and generally involve the evaluation of records from a recently conducted or an active study. According to the FDA's *Compliance Program Guidance Manual* for the monitoring of clinical sponsor compliance, the field inspector must evaluate at least six elements of a clinical study: (1) the selection of and directions to a monitor; (2) test article accountability; (3) assurance of IRB approval; (4) the adequacy of facilities; (5) continuing evaluation of data; and (6) records retention.

Monitor Selection and Directions   In this aspect of the inspection, the FDA investigator determines whether the clinical monitor is adequately qualified and whether the sponsor has given the monitor sufficient direction. Specifically, the inspector must determine:

• whether at least one individual has been charged with monitoring the progress of the investigation (if there are two or more monitors, the inspector must determine how the responsibilities are divided),

• what training, education, and experience qualify the monitor to oversee the progress of the clinical investigation;

- whether written procedures in addition to the protocol have been developed for monitoring the clinical investigations; and

- whether the sponsor has assured that the monitor has met his or her obligations.

Test Article Accountability   The inspector must determine whether the sponsor maintained adequate biologic accounting procedures before, during, and, if applicable, after a clinical investigation. Specifically, the inspector may be instructed to make as many as six separate determinations as to:

- whether the sponsor maintains accounting procedures for all usage of the test article, including records showing (1) the shipment dates, quantity, and serial, batch, lot or other identification number of units received; (2) the receipt dates and the quantity of returned articles; and (3) the names of investigators;

- whether the records are sufficient to allow a comparison of the investigator's test article usage against the amount shipped minus the amount returned;

- whether all unused or reusable supplies of the test article were returned to the sponsor when either: (1) the investigator discontinues, or finishes participating in, the clinical investigation; or (2) the investigation was terminated;

- the alternate disposition of the test article if all unused or reusable supplies of the test article were not returned to the sponsor, and how the sponsor determines the manner in which investigators account for unused or reusable supplies of the test article dispensed to subjects and not returned to the investigators;

- whether the alternate disposition was adequate to assure that humans or food-producing animals were not exposed to experimental risk; and

- whether records are maintained for alternate disposition of the test article.

Assurance of IRB Approval   For a clinical investigation subject to IRB approval, the FDA inspector must determine whether the sponsor maintains documentation showing that the clinical investigator obtained approval for the investigation before any human subjects were allowed to participate in the study.

Evaluating the Adequacy of Facilities   In this aspect of the inspection, the FDA inspector must determine whether the monitor assessed the adequacy of all facilities used by the study investigator—office, clinic, hospital.

Continuing Evaluation of Data   The inspector may examine records to evaluate the clinical sponsor's efficiency in reviewing data submitted by the investigator and in responding to reports of adverse reactions. Again, the inspector must make six determinations:

- whether all new case reports and other data received from the investigator regarding the safety of the test article are reviewed no later than ten working days after receipt;

- whether all case reports and other data received from the investigator are periodically evaluated for effectiveness as portions of the study are completed (included in this determination are the practices of the monitor);

- what actions are taken in response to incomplete case report forms;

- whether there is a system for tabulating the frequency and nature of adverse reactions;

- whether existing evidence indicates that the sponsor's present data receipt system is operating satisfactorily; and

- whether any deaths occurred among study subjects, and, if deaths did occur, what actions were taken to determine whether deaths were related to the use of the test article.

Record Retention   Finally, the inspector must determine the sponsor's compliance with record retention requirements. Sponsors must keep specified records and reports for two years after a biologic's marketing approval. If the product is not approved, the sponsor must maintain the records and reports for two years after shipment and delivery of the biologic for investigational use is discontinued and the FDA has been so notified.

This chapter reflects the author's assessment of CBER's Bioresearch Monitoring Program and is not intended to represent the official position of the FDA.

Chapter 14:
Biologics and the Regulation of Combination Products

by Michael R. Hamrell, Ph.D.
Chief, Regulatory Affairs Section*
Pharmaceutical and Regulatory Affairs Branch
Division of AIDS
National Institute of Allergy and Infectious Diseases
National Institutes of Health

The use of biological products in the treatment, prevention, and diagnosis of disease in man is a rapidly growing and diverse field. The explosion of technology in biological science has led to the development of an entirely new generation of products. Advances in gene splicing, cell manipulation, monoclonal antibodies, polymerase chain reaction (PCR) techniques, enzyme-linked immunosorbent assays (ELISA) and quantitative chromatographic assays have resulted in the development of scores of new products.

As these advances have been used to generate new products, however, they have created new challenges for the FDA. Advances such as the use of biological products in diagnostic testing have brought uncertainty regarding how, and by whom, such products should be regulated. Specifically, the advances have highlighted existing issues and created new regulatory and jurisdictional issues regarding three classes of FDA-regulated products: biologics, drugs, and medical devices.

* Dr. Hamrell is now the director of regulatory affairs at Vestar, Inc.

275

One complicating factor has been the constant change in the administrative and regulatory structure within the FDA. Biological products were regulated for many years by the Bureau of Biologics, which was merged with the Bureau of Drugs in 1982 to form the Center for Drugs and Biologics. In 1987, however, this center was split again into two major centers, the Center for Drug Evaluation and Research (CDER) and the Center for Biologics Evaluation and Research (CBER).

Similarly, for many years medical devices were regulated by the Bureau of Medical Devices or the Bureau of Radiological Health. In 1982, these two groups merged to form the Center for Devices and Radiological Health (CDRH).

These reorganizations have contributed to what some say is a confusing set of product classification precedents that are not always logical for advanced products. Complicating this further is the fact that, while one FDA center may have legal or historical jurisdiction over a product, the medical and technical expertise necessary for the review of that product may reside elsewhere within the agency.

Defining Biologics, Drugs, and Devices

Any analysis of product jurisdictional issues must begin with a discussion of the product definitions offered by FDA regulations. Again, technological advances tend to highlight some of the inherent limitations of these definitions.

Under federal regulations, CBER is given the authority to regulate all biological products. A biological product is defined as "any virus, therapeutic serum, toxin, antitoxin, or analogous product applicable to the prevention, treatment or cure of disease or injuries to man." This definition includes the use of a biologic as an aid in diagnosis or evaluation if the diagnostic substance involves a biologic reagent. To bring them under FDA jurisdiction, biological products also are legally defined as drugs or devices and are thus subject to all of the adulteration, misbranding, and registration provisions of the Food, Drug and Cosmetic Act (FD&C Act). (For more information on the legal and regulatory definitions of biologics, see Chapter 1.)

Medical devices, on the other hand, include any "instrument, apparatus, implement, machine, contrivance, implant, *in vitro* reagent or other similar or related article ... intended for use in the diagnosis of disease or other condition, or in the cure, mitigation, treatment, or prevention of disease ... or intended to affect the structure or any function of the body ... which does not achieve any of its principal intended purposes through chemical action within or on the body ... and which is not dependent upon being metabolized for achievement of any of its principal intended purposes."

Devices are regulated under section 201(h) of the FD&C Act, as amended by the Medical Device Amendments of 1976 and the Safe Medical Device Act of 1990. They include *in vitro* diagnostic products, which can be reagents, instruments, or systems used in the diagnosis of disease in order to provide treatment. Products used in the collection, processing and examination of specimens taken from the body for diagnostic purposes can also be classified as a medical device.

Drugs are regulated under section 505 of the FD&C Act. A drug is defined as any "article intended for use in the diagnosis, cure, mitigation, treatment or prevention of disease in man...." Included in the definition of drugs are articles that affect the structure or any function of the body of man.

If these definitions are taken literally, it is clear how they can lead to confusion regarding whether certain products are drugs, devices or biological products. And if the product is a combination of a drug, device and/or biologic, then the issue is even more confusing.

Combination Products

The incorporation of biological products into the realm of diagnostic testing has brought considerable confusion over whether such products should be classified as medical devices or biological products (or even drugs). For example, *in vitro* diagnostics (IVD) are medical devices used to collect, prepare, and examine specimens taken from the human body for diagnosis, treatment or prevention of disease. They can include reagents, instruments, culture media, and biological products used in such diagnostic products.

Such combination products have long been a controversial and sensitive subject for the FDA. For many years, FDA centers disagreed about who should regulate certain products. In some cases, the review and regulation of these combination products were divided or shared between the centers. Unfortunately, such situations occasionally lead to delays in product approval.

Combination products have been regulated by the different centers on a case-by-case basis, a situation that led to some degree of inconsistency. Later,

Memoranda of Understanding (MOU) were written by the centers to specify how they would regulate specific combination products that crossed jurisdictional lines. Other products not covered by the MOU often were handled by unwritten understandings or negotiations between the centers.

SMDA of 1990

In 1990, the U.S. Congress passed the Safe Medical Device Act (SMDA) of 1990. Although primarily concerned with the safety and regulation of medical devices, the legislation addressed the issue of combination products. Specifically, the law required the FDA to develop and implement regulations and procedures to deal with combination products, to determine the lead center for review in specific cases, and to help resolve some of the lingering issues regarding products whose classifications were ambiguous or unclear. Section 16 of the SMDA specifically required the agency to designate a component of FDA to have primary responsibility and jurisdiction for the premarket review and regulation of a product that was a combination of a drug, device, or biological product.

The Legislation and Related Regulations

To clarify what constitutes a combination product, the SMDA modified, in subtle ways, the definitions of drugs and devices. The changes made the definitions more specific and relevant to today's technology and helped resolve some of the product class confusion. Because advances in technology have blurred the traditional distinctions between some products, the changes were designed to ensure that the relevant product definitions were not mutually exclusive.

The SMDA's provisions required the FDA to establish formal procedures for determining the product categories for combination and difficult-to-classify products. On November 21, 1991, the FDA published final regulations outlining how the agency would decide whether such products were drugs, biologics, or medical devices. According to the regulation, this decision is to be based on the "primary mode of action" of the product or combination product. If the primary mode of action is that of a biologic, for example, CBER would have primary jurisdiction.

The regulation does not specify how the agency is to determine a product's primary mode of action. The FDA is permitted to make this determination as part of the process.

Under the regulation, the agency may use whatever resources are necessary to ensure a product's safety and effectiveness. After a center is given primary jurisdiction for a product, for example, the regulations permit consultation between centers, the call for a second application, or the separation of the original submission into distinct applications for each product component.

For scientific or administrative reasons, the designated lead center is allowed, with concurrence, to transfer the responsibility for a specific product to another center. If a product is assigned to a particular center, that center is responsible for regulating the product and evaluating product quality.

The regulation applies to the two different categories of products that have proven most problematic for the FDA: (1) any product that is a combination of a drug, device, or biological product under the FD&C Act; and (2) any drug, device, or biological product whose primary jurisdiction for premarket review and approval is unclear or disputed. The regulation does not apply to foods, veterinary products, cosmetics, or combinations of these products.

A combination product is defined as a product composed of two or more regulated components (i.e., drug/device, biologic/device, etc.) that are combined or mixed and produced as a single entity. This includes two or more separate products that are formulated and packaged as a single product as well as two or more distinct products that are packed separately within a single package.

A combination product also may be a drug, device, or biological product that is packaged separately, but according to its investigational plan or approved labeling is intended for use only with an individually specified investigational or approved drug, device, or biologic. However, the definition excludes the concomitant use of drugs, devices and biological products. It also excludes most products comprising two or more drugs, two or more devices, or two or more biologics.

In an accompanying regulation, the FDA delegated premarket approval authorities for combination products to directors, deputy directors, and certain other key supervisory personnel in CDER, CBER, and CDRH. The regulation grants these personnel reciprocal premarket approval authority to approve combination products, and delegates to the center with primary jurisdiction all authority necessary for premarket approval of a combination product. Thus, if CBER has primary jurisdiction over a device-biological combination product, CBER personnel have absolute authority to approve the product.

The Next Step: The FDA's Intercenter Agreements

While the SMDA and accompanying regulations address general jurisdictional and procedural issues regarding combination and difficult-to-classify products, they do not identify specific products or delineate how the products would be classified. To provide this guidance, CBER, CDER, and CDRH entered into written agreements:

• Intercenter Agreement Between CBER and CDRH;

• Intercenter Agreement Between CBER and CDER; and

• Intercenter Agreement Between CDER and CDRH.

Signed in October 1991 and issued as part of the SMDA's implementation, the agreements establish, by center, the allocation of responsibility for specific products and product categories. In formally documenting the types of products to be regulated by each center, the agreements resolve some of the long-standing confusion concerning the regulatory responsibility for certain product classes.

The agreements are publicly available, and sponsors seeking to investigate and eventually market a combination product should consult these documents. If a particular product is not covered by the guidance document, or if the center with primary jurisdiction is unclear, the sponsor may request a ruling on the product as outlined in the regulations.

Under the three agreements, intercenter jurisdictional committees were formed to review all jurisdictional questions. These committees, which meet on an *ad hoc* basis to discuss problem areas, will be a key to combination product regulation in the short term. Because these committees interpret the agreements, they will set the precedent for future products and jurisdictional issues.

The Intercenter Agreements and Product Designations

As their titles indicate, the intercenter agreements address three general types of difficult-to-classify or combination products:

- products that combine a biologic and a device or that could seemingly be designated in either product category;

- products that combine a biologic and a drug or that could seemingly be designated in either product category; and

- products that combine a drug and device or that could seemingly be designated in either product category;

The discussions below profile the two intercenter agreements affecting biologics and combination products that incorporate biological products.

The CBER-CDRH Agreement: Biologics and Devices

Generally, CBER will take the lead in regulating products that are medical devices utilized in or indicated for the collection, processing, or administration of biological products. CBER will be responsible for ensuring the safety and effectiveness of such products using its authorities under the Public Health Service Act and the FD&C Act.

CBER also will have lead responsibility for regulating medical devices used or indicated for the collection, processing, storage, or administration of blood products, blood components or analogous products as well as screening or confirmatory clinical laboratory tests associated with blood banking practices. This includes equipment designed to manipulate a patient's cells and return them to the patient's body.

In addition, CBER will have responsibility for regulating all *in vitro* tests and other medical devices indicated for the virus that causes AIDS (HIV), and for other retroviruses. This includes devices for collection, specimen containers, test kit components or support materials used or indicated for inactivation of these viruses (for a listing of some of the difficult-to-classify and combination biological-medical device products that will be regulated by CBER, see exhibit below).

Types of Devices That Are Designated as Biologics

- Devices used in the collection, processing or administration of a licensed biological product;
- *In vitro* reagents used to process biologics;
- *In vitro* reagents used as quality assurance reagents for a licensed biologic;
- Devices used in any aspect of blood bank-related biological products;
- Devices or *in vitro* reagents used in tissue typing;
- Devices intended as delivery systems for a biologic, packaged with the biologic;
- Devices filled with a licensed biologic as part of the manufacturing process and part of the final container or delivery system;
- Devices intended to serve as *in vivo* delivery systems for biological products;
- Cell separators and other blood processing equipment used to separate components for the production of blood products and biologics;
- Irradiating equipment used for the in-process inactivation of the Human Immunodeficiency Virus and other pathogens in biological products;
- Cellular and tissue implants; and
- Certain generic devices listed in 21 CFR §864.

CDRH will regulate all medical devices that are not categorically or specifically assigned to CBER. There are some biological products that, by agreement, are to be regulated as medical devices with the primary approval authority to be handled by CDRH. Additionally, if a medical device is to be used for therapeutic purposes, then CDRH will have lead responsibility (see exhibit below).

Biological Products Designated as Medical Devices

- Cultured skin;
- Tissue processing equipment and solutions used for transporting, storing or processing tissues and organs (except for items related to biological products);
- *In vitro* reagents used for the detection of infectious agents when used for diagnostic use;
- Cell separators and blood processing equipment used for therapeutic purposes;
- Irradiators intended for use in the inactivation of immunologically active cells from blood; and
- Nonlicensed human source materials from blood used in the manufacture of unlicensed diagnostic devices.

For biological products regulated as medical devices, both CBER and CDRH will share responsibility for administrative and enforcement activities related to:

- surveillance and compliance action;
- warning letters, seizures, injunctions, and prosecutions;
- promulgation of performance standards;
- FDA-requested and firm-initiated recalls;
- exemptions and variances from, and the application of, GMP regulations; and
- requests for export approval.

CDRH will continue to independently administer programs related to:

- small business assistance;
- registration and listing of firms;
- color additives;
- the GMP Advisory Committee; and
- all medical device reporting.

The CBER-CDER Intercenter Agreement: Biologics and Drugs

In their intercenter agreement, CBER and CDER have identified certain types of combination and difficult-to-classify products, and have designated the center that will play the lead role in regulating these products (see exhibit below).

CDER has responsibility for all classes of drugs, antibiotic products (excluding vaccines and allergenics), and chemically synthesized molecules. In general, CDER also will regulate all products whose main components' primary mode of action is that of a "drug."

Biologic and Drug Product Designations

Product Type	Center
Biological components used as a mode of localization or to affect the distribution of a drug.	CDER
Biological products used in combination to enhance effectiveness or ameliorate the toxicity of a drug product.	CDER
Biological products that have been combined with a radioactive component.	CBER
Combination products that consist of a biological component used as a mode of localization and a toxin component that is not a drug (e.g., ricin A toxin).	CBER
Combination products in which the drug component enhances the effectiveness or ameliorates the toxicity of the biological product.	CBER

Conversely, products or components of a combination exhibiting a primary action characteristic of a "biologic" will be regulated by CBER. As described above, CBER is responsible for all biological products subject to licensure, including vaccines, blood-derived products, immunoglobulin products, and products derived from human or animal tissue.

CDER and CBER also have agreed to apportion the scientific review of some products based on existing programs and the location of clinical and scientific expertise. According to the intercenter agreement, for example, CBER would handle the clinical data review component of collaborative reviews for products "intended for or acting by a mechanism of:

1. replacing the O_2 carrying function of red blood cells;

2. directly and specifically activating the proliferation of hematopoietic cells;

3. replacing plasma coagulation factors;

4. achieving a passive immune response;

5. diagnosing allergy or delayed type hypersensitivity by *in vivo* testing; and

6. inducing an antigen-specific active immune response or tolerant state (i.e., vaccines and allergenics) but excepting vaccines intended to induce a contraceptive state."

Such arrangements will facilitate the efficient utilization of existing resources and provide for rational program planning. However, either center may decline to participate in a requested collaborative review if expertise or resources are not available.

Product Jurisdiction Officer

Dealing with single or combination products that did not fit into predefined categories and products that had an unclear primary mode of action was a recurring issue during the development of the FDA's regulations. To address such situations, the regulation authorizes the FDA to designate a "product jurisdiction officer" to determine which component of the agency has primary jurisdiction over the product in question. In the regulations, the FDA ombudsman—a post in the FDA's Office of the Commissioner—is designated as the product jurisdiction officer.

The regulations provide extremely specific procedures for requesting a designation ruling for a product. Typically, the sponsor should file a request when the agency component responsible for review of its single or combination product is unclear or is not defined in the intercenter agreements. The sponsor should file the request before submitting any application for premarket review.

The designation request should contain all information necessary for the FDA to make a determination of the product's primary mode of action. If a sponsor wants to recommend to the agency that a specific center be given primary jurisdiction, the company should cite the reasons for its recommendation with the request.

The product jurisdiction officer will review and act on each request in writing within 15 days. The letter of designation will constitute the official agency determination for the product in question. Only under certain circumstances may this designation be changed.

As of late 1992, the FDA's ombudsman had been involved in jurisdictional decisions for 24 products. While 21 of these requests involved drug-device combinations, only three involved biologic-device combinations. Because product designation decisions involve confidential, proprietary information, the FDA has no plans to release highly specific information about the letters of designation filed or the agency's ultimate determinations.

Chapter 15:

CBER and Computer-Assisted Product License Applications

by Larry McCurdy
Computer Specialist
Center for Biologics Evaluation and Research
Food and Drug Administration

In the late 1980s and early 1990s, CBER faced pressures similar to those confronting other centers within the FDA. The increasing number, size, and complexity of IND and PLA submissions and a corresponding growth in average biologics review times made the need for computer-based solutions more acute.

In the final months of 1991, CBER moved to ensure that computer-assisted product license applications (CAPLA) would become in the 1990s what they had not in the 1980s, when only one CAPLA had been submitted to CBER. Center officials decided to adopt a systems-based approach to computerized submissions and began to assemble an information systems group to turn its vision into a reality.

A Short History of CAPLAs

With no CAPLA submissions through 1988, CBER had the opportunity to watch and learn from CDER's experiences with computer-assisted new drug applications (CANDA). One principal conclusion reached by center officials: CBER should provide some framework for CAPLA submissions rather than leave sponsors and reviewers to negotiate new computer hardware and software configurations for each submission.

Although CBER would provide companies with guidance shortly thereafter, it did not do so before the first CAPLA submission. In 1989, Genentech submitted to the Division of Cytokine Biology what is considered the first CAPLA. In a filing for gamma interferon, the company submitted narrative and summary reports, line listings, SAS data sets, and clinical data tables and summaries on diskettes.

Shortly after the Genentech submission, CBER formed its CAPLA Steering Committee, which consisted of senior managers and reviewers within the center. While this committee is now all but defunct, it succeeded in publishing a document entitled *Points to Consider: Computer Assisted Submissions for License Applications* in July 1990. That document was brief and basic, but did establish WordPerfect as the word processing package of choice and 3.5-inch diskettes as the medium of choice for electronic submissions.

CBER Shifts to "Systems-Based" Approach

In late 1991, CBER's approach to CAPLAs took a major turn. The center decided to adopt a systems-based approach to computerized submissions. In other words, it decided to develop a single, "unified" system that would require CBER reviewers to be trained only on that one system, which could accommodate the electronic submission of text, data, images, pictures, charts, and other forms of information.

By year-end 1991, CBER had developed a set of CAPLA objectives and a two- to three-year plan for fulfilling them. While improving and shortening PLA reviews is the core of any electronic application agenda, a set of eight objectives was established for the CAPLA program:

- Improve the quality of reviews.

- Shorten the review cycle.

- Simplify data retrieval and report generation.

- Improve the management of the approval process.

- Establish technical standards.

- Reduce paper storage and handling.

- Reduce the manufacturer's cost for preparing applications and increase U.S. competitiveness.

- Improve security.

Track 1 and CAPLAs

To accomplish these objectives, CBER has devised a two-phase, or "track," plan for CAPLA development. The first—Track 1—is a short-term approach designed to allow CBER to accept and review CAPLAs within the framework of the technologies and resources currently available at the center.

The most important factor shaping the nature of Track 1 was CBER's computing environment, which, in mid-1992, consisted of the following principal elements:

1. CBER has one main VAX 6510 computer serving the entire center, a VAX 6410 dedicated to the National Vaccine Program, and another VAX dedicated to CAPLAs.

2. CBER has a functioning wide area network (WAN) using Ethernet, T1 lines, DEC 3100 or 4200 file servers at each site, 3Comm, DECnet, TCP/IP, and PathWorks.

3. CBER's network uses DEC's VMS operating system, the office automation system ALL-IN-1, and Keyword's KEYpak, a document transfer and conversion utility integrated with ALL-IN-1. The acquisition of KEYpak was CBER's solution to the microcomputer diversity within the center. Specifically, the product allows reviewers to view text in the word processing formats they prefer, regardless of the formats in which the data are submitted or the formats their fellow reviewers are using.

4. CBER's desktop computers, which include both IBMs and Macintoshes, are individually connected to the network. All CBER staffers have personal computers.

5. Although WordPerfect has been the word processing system supported by CBER and the CAPLA program, word processing formats are now somewhat less critical, since CBER's recently acquired document conversion package automatically converts information into the reviewer's preferred format. Therefore, sponsors are no longer limited to submitting text in WordPerfect, but may also use WPS, WPS-Plus, and Microsoft Word for either the IBM PC or the Macintosh.

The Track 1 plan will not impose inflexible standards on CAPLA sponsors. In fact, because Track 1's objective is to permit electronic submissions and reviews within the framework of the various hardware and software now being used by CBER reviewers, the plan had to be extremely flexible.

Track 1 also is a relatively straightforward approach that does not require major capital investment by participating sponsors. In fact, CBER's assumption in Track 1 is that CAPLAs will principally involve the submission of text and will be filed on diskettes. Just as importantly, because the approach involves the use of CBER equipment and resources, sponsors are not required to develop and outfit FDA reviewers with costly hardware and software packages.

Track 1: How It Will Work Essentially there will be two types of Track 1 CAPLAs. In the first, the reviewers will use the sponsor's diskettes on their personal computers and will share files by physically delivering the diskettes to other reviewers.

For the second type of Track 1 CAPLA, the files stored on the submitted diskettes will be loaded into a central repository on CBER's VAX computer. Reviewers will access the files through the center's wide area network. When a file is accessed, the file transfer and conversion package automatically converts the file to the reviewer's preferred word processing format. The system will not permit reviewers to modify the copy of the file on the central repository. Once a copy of the file is transferred to the reviewer's system, however, that copy may be manipulated (e.g., cut-and-paste, re-analysis) by the reviewer.

Track 1 offers several advantages to CBER reviewers:

- immediate access to submission files at any site via the reviewer's desktop terminal;

- simultaneous file access for multiple reviewers;

- the ability to circulate documents electronically over the network for comment by other reviewers without the original reviewer having to leave his or her desk;

- the elimination of the need to reformat documents for different hardware or word processors because KEYpak automatically handles file conversions;

- enhanced security, because there is no potential for lost or misplaced volumes or disks; and

- the ability for reviewers to work at home, provided they have the necessary equipment.

The first Track 1 pilot project was initiated in late 1991. The pilot is a Connaught Laboratories CAPLA that consists of an ELA in a WordPerfect file. Unfortunately, KEYpak was not available to CBER at the time of the CAPLA submission. Because the ELA was generated in a Macintosh environment, the text of the application had to be converted to WordPerfect by a consultant to Connaught Laboratories. The pilot is being evaluated as it progresses.

While CBER views Track 1 as a short-term solution, the center will continue to offer the approach as an option to those companies wanting to use it. Even after the center develops and implements a more comprehensive and advanced approach, CBER believes that the flexibility and simplicity of Track 1 will continue to appeal to many firms, particularly smaller, less experienced companies.

Track 2: The Future of CAPLAs

Although Track 2 planning is still in progress, it is during this second phase that CBER hopes that CAPLAs will reach their full potential. Currently, the center is evaluating more than a dozen different hardware, software, and systems elements for possible inclusion in the Track 2 system.

CBER has tentatively settled on a general concept for the Track 2 system (see exhibit below). At the core of the system are likely to be four or five heterogeneous databases, each storing different types of information typically submitted in PLAs and ELAs:

Text Databases. The system will include two text databases, one for the narrative portion of the PLA or ELA and another for reviewer annotations. The narrative text database will incorporate sophisticated search and retrieval capabilities, including the ability to locate all references of a phrase within a submission, and capabilities for cross-referencing with other submissions. The location of references may be based on keywords or full text search. Reviewer annotations may be made electronically in the margin of the narrative, and the annotations of several reviewers can be consolidated later into a new document that can be sent electronically via MCI mail to the sponsor.

Track 2 Design Principles

- Based on a single system approach throughout CBER.

- All reviewers will have access from their desktop computer to all submissions via a network.

- Same interface for all reviewers.

- Built upon a toolbox approach - several software packages are bundled together to offer a wide variety of features (such as text search, annotations, etc.) to aid reviewers in manipulating documents and data.

- Supports text, data, graphics, and image information.

- Reference documents and databases available on-line.

- Case report forms and other source data available on-line.

- Interface will support access to multiple types of information simultaneously via windows.

- Based on electronic publishing as the document handling tool.

- Supports rapid, electronic navigation through a document via the table of contents, index, and other references.

- Runs on the reviewer's desktop computer of choice, either Macintosh, PC, or VAX station.

- Supports variability of review methodology.

- Supports management of the approval process.

- Supports variability among specific elements of individual submissions.

- Automatic evaluation of completeness of submission.

- Improved communication with sponsor.

- Data interchange between the sponsor and the FDA is hardware- and software-independent.

- Standards will be developed or adopted and used wherever possible.

Table Database. A table database will have the ability to integrate spread-sheets and statistical packages so CBER reviewers can manipulate data. The CAPLA system automatically will populate the spreadsheets and statistical packages so CBER reviewers must no longer rekey data. This database will allow reviewers to verify results and manipulate data. All data will be accessible from the screen to provide "on-screen review" capability.

Image Database. The image database will permit the storage of pictures, gels, charts, X-rays, blueprints, high pressure liquid chromatography (HPLC), handwritten lab notes, case report forms, and other information that can be stored as images.

Relational Database. A relational database will contain general information about each submission.

These heterogeneous databases will be linked to form what might be termed a "virtual document." This means that, as a reviewer moves through the submission electronically, the system automatically will pull information from the respective databases in the order in which it appears in the hardcopy format. In other words, if a reviewer is reading a narrative summary in the text database and there is a table that appears in the summary, the system automatically will move to the table database, present the table on the screen, and then shift back to the text database as the reviewer advances through the document. CBER plans to use a graphical user interface (GUI) to provide reviewers with a single computing environment that will simplify access to all parts of the CAPLA system.

Because it can support heterogeneous database access and hyperlinks, electronic publishing appears to be the software platform best suited for the Track 2 CAPLA. In fact, the existing generation of some electronic publishing packages provides virtually all of the features required by the project.

All CAPLA submission files will be stored in a central repository maintained within the Parklawn Computer Center, the FDA's principal computing facility and service center. This central facility will serve as a single entry point for submissions sent to the FDA, where the agency will perform computer checks for conformance to PLA format and completeness before downloading working files to sites that CBER reviewers occupy.

Storing the submissions in a central repository will provide a high level of physical security. The agency will maintain audit trails for each access to a file, including what operations are performed (i.e., read, modify, delete, print, etc.), what software was invoked, and how long each software application was employed. The data on the duration of access times will assist CBER management in assessing time frames for certain types of reviews.

Track 2: A Look Ahead

Despite the groundwork completed to date, critical Track 2 design elements remain (see exhibit below), particularly the development of technical and data exchange standards. CBER must establish technical standards for naming and numbering conventions, data formats, file formats, document structure, and media. The center will coordinate these with industry to streamline the production, transmittal, and review of submissions. Work on these standards is expected to constitute the single largest effort in the CAPLA project.

CBER also will study various means of implementing standards for the data exchange interface between sponsors and the FDA. Special attention will be placed on existing international standards, such as the Standard Generalized Markup Language (SGML), to identify an interface that is hardware and software independent.

While still in the Track 2 systems analysis stage, CBER remains conservative in its projections. Ideally, the center hopes to have a Track 2 system with 80 percent functionality by early 1995. By late 1993, however, CBER hopes to work with a manufacturer on a Track 2 pilot project, when the center hopes to have 40 percent of Track 2 functionality available.

This chapter reflects the author's assessment of CBER's CAPLA program and is not intended to represent the official position of the FDA.

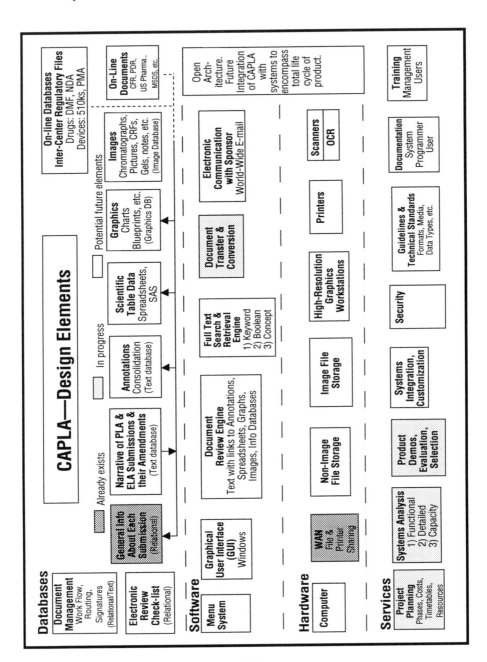

295

Appendix 1

Guidelines Available for Human Biological Products

Interpretive Guidelines of the Additional Standards for Source Plasma (Human) (September 1973).

Reviewing Amendments to Include Plasmapheresis of Hemophiliacs (July 1976).

Immune Serum Globulin (Human) (March 1978).

Interpretation of Potency Test Results for All Forms of Adsorbed Diphtheria and Tetanus Toxoids (April 1979).

Immunization of Source Plasma (Human) Donors with Blood Substances (June 1980).

Collection of Human Leukocytes for Further Manufacturing (January 1981).

Platelet Testing - Approval of New Procedures and Equipment (July 1981).

Adding Heparin to Empty Containers for Collection of Heparinized Source Plasma (Human) (August 1981).

Release of Pneumococcal Vaccine, Polyvalent (January 1982).

Meningococcal Polysaccharide Vaccines (July 1985).

Uniform Labeling of Blood and Blood Components (August 1985).

Submitting Documentation for Packaging for Human Drugs and Biologics (February 1987).

Submitting Documentation for the Stability of Human Drugs and Biologics (February 1987).*

Sterile Drug Products Produced by Aseptic Processing (June 1987).

Guideline on General Principles of Process Validation (May 1987).*

Validation of the Limulus Amebocyte Lysate Test (December 1987).*

Collection of Platelets, Pheresis (October 1988).

Design of Clinical Trials for Evaluation of Safety and Efficacy of Allergenic Products for Therapeutic Uses (November 1988).

Collection of Blood or Blood Products From Donors With Positive Tests for Infectious Disease Markers ("High Risk" Donors) (September 1989).

Test for Residual Moisture for Biological Products (January 1990).

Draft Guideline for Reporting-ABR's-Adverse Reactions to Licensed Biological Products (March 1990).

Draft Guideline for Quality Assurance in Blood Establishments (June 1993).

Guidance on Alternatives to Lot Release for Licensed Biological Products (July 1993).

Interim Guidance on Applicability of User Fees to: (1) Applications Withdrawn Before Filing Decision, or (2) Applications the Agency Has Refused to File and That Are Resubmitted or Filed Over Protest (July 1993).

Interim Guidance on Separate Marketing Applications and Clinical Data for Purposes of Assessing User Fees Under the Prescription Drug User Fee Act of 1992 (July 1993).

Center for Biologics Evaluation and Research (CBER): Refusal To File (RTF) Guidance for Product License Applications (PLAs) and Establishment License Applications (ELAs) (July 1993).

Interim Guidance Document for Waivers Of and Reductions In User Fees (July 1993).

Office of Establishment Licensing and Product Surveillance Advertising and Promotional Labeling Staff (July 1993).

Points to Consider (PTC) Available for Human Biological Products

PTC in the Manufacture of *In Vitro* Monoclonal Antibody Products Subject to Licensure (June 1983).*

PTC in the Production and Testing of Interferon Intended for Investigational Use in Humans (July 1983).*

PTC in the Production and Testing of New Drugs and Biologicals Produced by Recombinant DNA Technology (April 1985).*

PTC in the Manufacture and Testing of Monoclonal Antibody Products for Human Use (June 1987, revised document expected in late 1993).*

PTC in the Collection, Processing and Testing of *Ex-Vivo*-Activated Mononuclear Leukocytes for Administration to Humans (August 1989).*

PTC in the Manufacture and Clinical Evaluation of *In Vitro* Tests to Detect Antibodies to the Human Immunodeficiency Virus, Type I (draft) (August 1989).

PTC: Computer Assisted Submissions for License Applications (July 1990).

PTC in the Safety Evaluation of Hemoglobin-Based Oxygen Carriers (August 1990).*

Cytokine and Growth Factor Pre-Pivotal Trial Information Package (April 1990).

PTC in Human Somatic Cell Therapy and Gene Therapy (August 1991).*

PTC in the Manufacture of *In Vitro* Monoclonal Antibody Products for Further Manufacturing into Blood Grouping Reagent and Anti-Human Globulin (March 1992).*

PTC in the Design and Implementation of Field Trials for Blood Grouping Reagents and Anti-Human Globulin (March 1992).

Supplement to the Points to Consider in the Production and Testing of New Drugs and Biologicals Produced by Recombinant DNA Technology: Nucleic Acid Characterization and Genetic Stability (April 1992).*

PTC in the Characterization of Cell Lines Used to Produce Biologicals (May 1993).*

* Includes guidance on preclinical testing issues.

Others

Interstate Shipment of Interferon for Investigational Use in Laboratory Research Animals or Tests *In Vitro* (November 1983).

Recommended Methods for Short Ragweed Pollen Extracts (November 1985).

Information Relevant to the Manufacture of Acellular Pertussis Vaccine (August 1989).

Recommended Methods for Evaluating Potency, Specificity, and Reactivity of Anti-Human Globulin (March 1992).

Recommended Methods for Blood Grouping Reagents Evaluation (March 1992).

FDA's Policy Statement Concerning Manufacturing Arrangements for Licensed Biologicals (November 1992).

Memoranda and Related Documents Pertaining to Human Blood and Blood Products

Revision of October 7, 1988 Memorandum Concerning Red Blood Cell Immunization Programs (December 1992).

Volume Limits for Automated Collection of Source Plasma (November 1992).

Nomenclature for Monoclonal Blood Grouping Reagents (September 1992).

Changes in Equipment for Processing Blood Donor Samples (July 1992).

Revised Recommendations for the Prevention of Human Immunodeficiency Virus (HIV) Transmission by Blood and Blood Products (April 1992).

Use of Fluorognost HIV-1 Immunofluorescent Assay (IFA) (April 1992).

Revised Recommendations for Testing Whole Blood, Blood Components, Source Plasma and Source Leukocytes for Antibody to Hepatitis C Virus Encoded Antigen (Anti-HCV) (April 1992).

Exemptions to Permit Persons with a History of Viral Hepatitis Before the Age of Eleven Years to Serve as Donors of Whole Blood and Plasma: Alternative Procedures, 21 CFR 640.120 (April 1992).

Clarification of FDA Recommendations for Donor Deferral and Product Distribution Based on the Results of Syphilis Testing (December 1991).

Disposition of Blood Products Intended for Autologous Use That Test Repeatedly Reactive for Anti-HCV (September 1991).

FDA Recommendations Concerning Testing for Antibody to Hepatitis B Core Antigen (Anti-HBc) (September 1991).

Revision to 26 October 1989 Guideline for Collection of Blood or Blood Products from Donors with Positive Tests for Infectious Disease Markers ("High Risk" Donors) (April 1991).

Responsibilities of Blood Establishments Related to Errors & Accidents in the Manufacture of Blood Components (March 1991).

Deficiencies Relating to the Manufacture of Blood and Blood Components (March 1991).

FDA Request for information on blood storage patterns and red cell contamination by *Yersinia enterocolitica* (March 1991).

Use of Genetic Systems HIV-2 EIA (June 1990).

Autologous Blood Collection and Processing Procedures (February 1990).

Guideline for Collection of Blood or Blood Products from Donors with Positive Tests for Infectious Disease Markers ("High Risk" Donors) (October 1989).

Abbott Laboratories' HIVAG-1™ test for HIV-1 antigen(s) not recommended for use as a donor screen (October 1989).

Requirements for Computerization of Blood Establishments (September 1989).

Use of Recombigen® HIV-1 Latex Agglutination (LA) Test (August 1989).

HTLV-I Antibody Testing (July 1989).

Guidance for Autologous Blood and Blood Components (March 1989).

Use of the Recombigen® HIV-1 LA Test (February 1989).

HTLV-I Antibody Testing (November 1988).

Antibody to Human T-Cell Lymphotropic Virus, Type I (HTLV-I) Release Panel (October 1988).

Recommendations for Changeover from Use of Fresh Immunizing Red Blood Cells to Use of Frozen Immunizing Red Blood Cells Stored a Minimum of Six Months Prior to Use (October 1988).

Guideline for the Collection of Platelets, Pheresis Prepared by Automated Methods (October 1988).

Criteria for Exemption of Lot Release (August 1988).

Physician Substitutes (August 1988).

Discontinuance of Pre-licensing Inspection for Immunization Using Licensed Tetanus Toxoid and Hepatitis B and Rabies Vaccines (July 1988).

Recommendations for Implementation of Computerization in Blood Establishments (April 1988).

Control of Unsuitable Blood and Blood Components (April 1988).

Handling of Human Blood Source Materials (December 1987).

Extension of Dating Period for Storage of Red Blood Cells, Frozen (December 1987).

Recommendations for the Management of Donors and Units that are Initially Reactive for Hepatitis B Surface Antigen (HBsAG) (December 1987).

Deferral of Donors Who Have Received Human Pituitary-Derived Growth Hormone (November 1987).

Reduction of Maximum Platelet Storage Period to 5 Days in an Approved Container (June 1986).

In Vitro Diagnostic Reagent Manufacturers (December 1985).

Equivalent Methods for Compatibility Testing (December 1984).

Plasma Derived from Therapeutic Plasma Exchange (December 1984).

Deferral of Blood Donors who have Received the Drug Accutane (isotretinoin/Roche; 13-cis-retinoic acid) (February 1984).

Requirements for Infrequent Plasmapheresis Donors (August 1982).

Guideline for Uniform Labeling of Blood and Blood Components (August 1985).

Appendix 2

Glossary of Acronyms

ADME	Absorption, Distribution, Metabolism and Excretion
BEPA	Division of Blood Establishment and Product Applications
BRM	Biological Response Modifier
CAPLA	Computer Assisted Product License Application
CDC	Center for Disease Control
CBER	Center for Biologics Evaluation and Research
CDER	Center for Drug Evaluation and Research
CDRH	Center for Devices and Radiological Health
CFR	Code of Federal Regulations
cGMP	current Good Manufacturing Practices
COI	Conflict of Interest
CPG	Compliance Policy Guides
CRO	Contract Research Organization
CSO	Consumer Safety Officer
DARP	Division of Application Review and Policy
DFO	Designated Federal Official
DHHS	Department of Health and Human Services
DVRPA	Division of Vaccine and Related Products Applications
ELA	Establishment License Application
FDA	Food and Drug Administration
FD&C	Food, Drug & Cosmetic Act
GCP	Good Clinical Practices

GLP	Good Laboratory Practice
GRAS	Generally Recognized As Safe
GUI	Graphical User Interface
IDE	Investigational Device Exemptions
IND	Investigational New Drug Application
IOM	Institute Of Medicine
IRB	Institutional Review Board
MCB	Master Cell Bank
MOV	Memoranda Of Understanding
MTD	Maximum Tolerated Dose
MWCB	Manufacturers Working Cell Bank
NDA	New Drug Application
NIH	National Institutes of Health
PLA	Product License Application
PMA	Pharmaceutical Manufacturers Association
QAU	Quality Assurance Unit
SGE	Special Government Employees
SGML	Standard Generalized Markup Language
SMDA	Safe Medical Device Act
SOP	Standard Operating Procedure
VARBPAC	Vaccines And Related Biological Products Advisory Committee

Index